New Trends of Powder Engineering and Additive Manufacturing (Editorial Board Members' Collection Series)

New Trends of Powder Engineering and Additive Manufacturing (Editorial Board Members' Collection Series)

Guest Editors

Joan-Josep Suñol
Leszek Adam Dobrzański

MDPI

Basel • Beijing • Wuhan • Barcelona • Belgrade • Novi Sad • Cluj • Manchester

Guest Editors
Joan-Josep Suñol
Department of Physics
University of Girona
Girona
Spain

Leszek Adam Dobrzański
ASKLEPIOS
Medical and Dental Engineering Centre
for Research, Design and Production
Gliwice
Poland

Editorial Office
MDPI AG
Grossgpeteranlage 5
4052 Basel, Switzerland

This is a reprint of the Special Issue, published open access by the journal *Metals* (ISSN 2075-4701), freely accessible at: https://www.mdpi.com/journal/metals/special_issues/0823RO6D1L.

For citation purposes, cite each article independently as indicated on the article page online and as indicated below:

Lastname, A.A.; Lastname, B.B. Article Title. *Journal Name* **Year**, *Volume Number*, Page Range.

ISBN 978-3-7258-7408-8 (Hbk)
ISBN 978-3-7258-7409-5 (PDF)
https://doi.org/10.3390/books978-3-7258-7409-5

Contents

About the Editors

Joan-Josep Suñol

Joan-Josep Suñol: Professor of Department of Physics, Campus Montilivi s/n, University of Girona, 17003 Girona, Spain. His research interests are powder metallurgy, structural analysis, thermal analysis, mechanical alloying, and nanocrystalline. He has more than 285 scientific publications to his name (Science Citation Index Expanded). Since 2015, he has made more than 80 publications (with 70% of articles in journals indexed with Q1 or Q2 impact factors), and his number of citations/year is greater than 340, making his h-index 32. One of his articles from 2014 has more than 1120 citations. He has presented more than 360 communications at scientific congresses. He is an Associate Editor of the indexed journals (Q1-Q2) *Journal of Thermal Analysis and Calorimetry* (since 2010), *Materials*, *Metals*, and *Chemistry Africa*. He is a scientific reviewer of articles for more than thirty indexed journals. He has experience as an evaluator of projects of ANEP, AECID, AGAUR, several Spanish Territorial agencies, and Polish, Austrian, Chilean, and Argentine agencies.

He has been responsible for the Research Group on Materials and Thermodynamics (GRMT) at the University of Girona since 2008. He was responsible for integrating our thermal analysis laboratory as a linked unit (Thermal Analysis Unit) into the technical research services of the university. Orcid: 0000-0002-7653-3415.

Leszek Adam Dobrzański

Leszek Adam Dobrzański, Prof., DSc, PhD, MSc, Eng, Hon. Prof., M Dr HC, is currently a full Professor and the director of the Science Center, as well as the chairman of the supervisory council of the Medical and Dental Engineering Center for Research, Design and Production ASKLEPIOS Ltd in Gliwice, Poland. His research interests are materials engineering, nanotechnology, biomaterials, medical, dental, manufacturing and surface engineering, machine building and automation, and management and organization. He is the president of the World Academy of Materials and Manufacturing Engineering and of the Association of Computational Materials Science and Surface Engineering. He is the Editor-in-Chief of the *Journal of Achievements in Materials and Manufacturing Engineering* and the *Archives of Materials Science and Engineering*, and from 1991 to 2017, he worked as a full Professor, dean of the faculty, and vice rector at the Silesian University of Technology in Gliwice, Poland. On October 1, 2024, he was granted the honorary status of Emeritus Professor of the Silesian University of Technology in Gliwice, Poland.

His scientific output includes over 3,000 primarily English-language publications, including approximately: 60 books; 60 chapters in books and monographs; 250 articles listed on the Web of Science; and 60 patents, utility models, and patent applications. He has held visiting professorships at almost 100 universities across all continents, given over 120 invited lectures at international scientific conferences in many countries worldwide, and organized over 100 of them as chairman of organizing committees. He is the sole author of the *Engineering Materials with the Fundamentals of Materials Processing Technologies*. He ranks in the middle of the top 2% of the world's most cited scientists according to a list compiled by Elsevier and Stanford University, USA. Researcher unique identifier: ORCID 0000-0002-7584-8917.

metals

MDPI

Editorial

New Trends in Powder Engineering and Additive Manufacturing (Editorial Board Members' Collection Series)

Leszek Adam Dobrzański [1] and Joan-Josep Suñol [2,*]

[1] Medical and Dental Engineering Centre for Research, Design and Production ASKLEPIOS, 44-100 Gliwice, Poland; leszek.adam@gmail.com
[2] P2, Department of Physics, Campus Montilivi s/n, University of Girona, 17003 Girona, Spain
* Correspondence: joanjosep.sunyol@udg.edu

1. Introduction and Scope

The production of metal alloys and compounds in powder form has traditionally been linked to techniques such as gas atomization or mechanical alloying. For the manufacture of bulk pieces and components, powder metallurgy routes based on the pressing and sintering of powders have been used. In recent decades, additive manufacturing techniques have appeared as an alternative to produce parts with complex geometry and reduce the loss of material. The integration of powder engineering and additive manufacturing has driven the development of advanced materials and more efficient processes. Designing powders with controlled characteristics allows for improved repeatability, reduced defects, and expanded additive manufacturing applications (in demanding industries such as aerospace, automotive, and biomedical). Together, both disciplines allow optimization from the raw material to the final product, enhancing manufacturing innovation.

This Special Issue "New Trends in Powder Engineering and Additive Manufacturing" focuses on powder metallurgy, which is a set of fabrication techniques related to three major processing steps. First, the precursor material is physically powdered (micro- or nanometric particles). Second, the powder is consolidated to obtain bulk specimens (traditionally by injection into a mold or passed through a dye). Third, pressure and/or temperature is applied. Powder metallurgy is now also applied in the production of composites. Furthermore, new topics have emerged, such as additive manufacturing, MA, the circular economy or raw materials.

2. Overview of the Contributions

A total of eight peer-reviewed articles were published as part of this Special Issue. They cover several of the key aspects of the powder engineering and additive manufacturing process, ranging from powders to AM bulk parts. Although an incomplete overview (only eight manuscripts), it provides information on some of the most interesting aspects and materials.

The articles cover several topics, such as the correlations between processing and the microstructure or between the microstructure and the mechanical properties, as well as the corrosion resistance.

Four articles analyze stainless steel, focusing on the mechanical response, corrosion, and wear resistance. Steels are a traditional topic in metal science and engineering, due to their use in many fields such as biomedical, construction, automotive, chemical, aerospace, marine, nuclear, and petrochemical.

One article analyzes the corrosion resistance of 316L stainless steel [1]. One pathway to improve the corrosion resistance of stainless steels is the addition of chromium (<12 weight

%). Its corrosion resistance in aggressive conditions favors its applicability in the nuclear, biomedical, marine, and petrochemical industries. The 316L stainless steel was produced via ball milling followed by spark plasma sintering, an additive manufacturing technique. This study shows that the reduction in the grain size negatively affects the electrochemical passivation behavior. However, the passivation in HNO_3 before consolidation may mitigate this effect. The corrosion resistance was improved by the prior passivation of the powder (especially in small particle sizes). On the other hand, the decrease in grain size provoked the reduction in the length of the passive domain.

In another work, additive manufacturing extrusion was applied to produce 316L stainless steel [2]. Different amounts of copper (Cu) powder were added to the paste composition to reduce the sintering temperature (by promoting persistent liquid phase sintering). The authors recognized that no adequate densification was achieved, which would be necessary to yield competitive properties. One feature of this study was the use of methylcellulose as an alternative binder to wax-based binders. The relative density and Vickers hardness increased with the Cu content.

Another work focused on 316 austenitic steels and nickel-based superalloys produced using laser powder bed fusion [3] to fabricate multi-material components. AM technology is a pathway for the development of Functionally Graded Materials (FGMs). One of the principal interests of additive manufacturing is the customization of components and parts according to the required geometry and properties. In this case, the authors discussed its applicability to the aerospace, energy, biomechanical, automotive, and marine fields. The analysis of the results confirmed the feasibility of producing FGM components at the layer level. The interface of the samples produced appeared to be free of defects such as lack of fusion and delamination, which is a signal of a good metallurgical bond.

The final work related to steels explored the technical feasibility of the isostatic hot pressing (HIP) fabrication of integral bimetallic components using supercritical carbon dioxide, turbomachinery steel powder, and nickel alloy [4]. There was a dense distribution of Ti-rich carbonitrides, and alumina particles were found to decorate prior particle boundaries near the interface on the 282 side, affecting both the tensile strength across the interface and the tensile failure location. A pilot-scale bimetallic SS415/282 pipe was developed to show the viability of the scale-up. In addition, the simulations performed with DICTRA software helped to explain the interdiffusion and phase evolution at the interface between steel and alloy 282 during the HIP cycle.

Three articles focused on Al-based alloys (Al-Mg-based, Al-Fe-based, Al-Zn-based). These alloys are used in the automotive, aerospace, and transport industries. They commonly have high strength, good plasticity, and high performance in corrosion resistance. Thus, it is expected that aluminum alloys will effectively reduce the self-weight of components, while maintaining sufficient strength and toughness.

In one of the articles, two Al-based alloys (Al-Fe-Si-Cr-Ni) were produced via gas atomization. This technique favors the formation of spherical particles that can be sieved and used in additive manufacturing processes. The authors recommend minimizing the internal porosity of the powders and discuss the influence of the alloying element content on the solidification processes [5]. Gas atomization has made it possible to increase the solid solubility of elements with a low diffusion coefficient (Fe, Ni, Mn, Zr, Cr) in the α-Al matrix. As a result, it has led to (a) the formation of metastable phases and (b) a higher performance of aluminum alloys at high temperatures.

Another work analyzed the high-temperature microstructural and mechanical properties of the Al-Mg-Sc-Zr alloy processed by the additive manufacturing technique of laser powder bed fusion (LPBF), highlighting its potential for aerospace applications [6]. The researchers also studied the microstructural evolution by highlighting the formation of

https://doi.org/10.3390/met16010060

nano-Al_3(Sc, Zr) particles that reinforce the alloy. The reinforcement mechanisms were heterogeneous nucleation and grain refinement. Complementary fracture tests revealed ductile behavior, with abundant dimples on the fracture surfaces, which was attributed to the second-phase particles.

In a third Al-based article, the Al-rich Al-Zn-Mg-Si alloy was produced via laser-directed energy deposition, DED [7]. In train railways, it is good to repair aluminum alloy through directional energy deposition in order to (a) reduce the cost and (b) improve the performance of the aluminum alloy. The analysis of the results showed low porosity and a correct bonding interface. The grains in the repaired area were columnar crystals growing vertically along the boundary of the melt pool. A sharp decline in tensile strength and hardness from the heat-affected zone to the matrix was detected, which was attributed to the large temperature gradient in the deposition process, consequently provoking the softening of the substrate. Due to the softening, an additional heat treatment was recommended to improve the mechanical response.

In the final study, zinc dendritic particles were produced via electrolysis from an alkaline electrolyte [8]. The formed dendrites showed a (002) preferred orientation, which was strong, as the precursors of the dendrites had a (101) (002) preferred orientation. This (002 plane) orientation was explained by the low surface energy. Likewise, zinc dendrites formed under diffusion control during electrolysis. At low overpotentials (−100 mV), irregular particles called dendrite precursors are generated. At higher overpotentials (−160 mV and above), dendrites of different shapes form, from compact and massive to branched in 2D and 3D. These could be applied in rechargeable batteries and catalysts.

3. Conclusions and Outlook

The research compiled in this Special Issue effectively demonstrates the dynamic and multifaceted nature of current advancements in powder engineering and additive manufacturing. The collected contributions underscore the increased technological and scientific relevance of steels, Al alloys, and Zn in terms of applications in different fields such as batteries, catalysis, aerospace, nuclear, biomedical, marine, petrochemical, and so on. These applications focus on the mechanical and functional properties as well as the corrosion and wear resistance.

The findings demonstrate that careful control of processing conditions and optimized microstructure favor significant property improvements. Together, the contributions gathered in this Special Issue offer valuable insight and experimental support for the design of robust, cost-effective, and high-performance pieces and components. This Special Issue can serve a reference for further academic and technological research on powder technology and additive manufacturing.

Author Contributions: Conceptualization, L.A.D. and J.-J.S.; methodology, L.A.D.; writing—original draft preparation, L.A.D. and J.-J.S.; writing—review and editing, J.-J.S.; All authors have read and agreed to the published version of the manuscript.

Funding: This research received no external funding.

Acknowledgments: As Guest Editors, we would like to extend our sincere gratitude to the Editorial staff for their dedicated support throughout the entire publication process.

Conflicts of Interest: The authors declare no conflicts of interest.

https://doi.org/10.3390/met16010060

References

1. Hug, E.; Keller, C.; Folton, C.; Papin, J.; Tabalaiev, K.; Marnier, G. Ultrafine Grain 316L Stainless Steel Manufactured by Ball Milling and Spark Plasma Sintering: Consequences on the Corrosion Resistance in Chloride Media. *Metals* **2024**, *14*, 864. [CrossRef]
2. Silvain, J.-F.; Gifford, D.L.; Fourcade, S.; Cuzacq, L.; Grosseau-Poussard, J.-L.; Debiemme-Chouvy, C.; Tessier Doyen, N.; Lu, Y. Study on Debinding and Sintering Conditions in Extrusion-Based Additive Manufacturing of 316L and 316L + Cu. *Metals* **2023**, *13*, 1858. [CrossRef]
3. Angelastro, A.; Posa, P.; Errico, V.; Campanelli, S.L. A Systematic Study on Layer-Level Multi-Material Fabrication of Parts via Laser-Powder Bed Fusion Process. *Metals* **2023**, *13*, 1588. [CrossRef]
4. Huang, S.; Shen, C.; Samarov, V. Processing, Microstructure, and Properties of Bimetallic Steel-Ni Alloy Powder HIP. *Metals* **2024**, *14*, 118. [CrossRef]
5. Bhatt, B.; Martucci, A.; Virgillito, E.; Gobber, F.; Bondioli, F.; Manfredi, D.; Lombardi, M.; Fino, P. Deciphering Microstructures and Phases of Gas-Atomised Novel Al-Fe-Si-Cr-Ni Alloys. *Metals* **2024**, *14*, 17. [CrossRef]
6. Ren, Z.; Zhang, H.; Shu, X.; Xu, H.; Chen, S.; Ding, Y.; Liang, L.; Qiu, Z.; Yang, Y.; Zheng, Y. Study on Microstructure and High-Temperature Mechanical Properties of Al-Mg-Sc-Zr Alloy Processed by LPBF. *Metals* **2024**, *14*, 890. [CrossRef]
7. Chen, J.; Yu, S.; Yang, J.; Xu, R.; Li, R.; Huang, S.; Zhu, H.; Liu, X. Research on the Microstructure and Mechanical Properties of Repaired 7N01 Aluminum Alloy by Laser-Directed Energy Deposition with Sc Modified Al-Zn-Mg. *Metals* **2023**, *13*, 829. [CrossRef]
8. Nikolić, N.D.; Lović, J.D.; Maksimović, V.M.; Vuković, N.S.; Ignjatović, N.L.; Živković, P.M.; Stevanović, S.I. Correlation Between Morphology and Crystal Structure of Electrolytically Produced Zinc Dendritic Particles. *Metals* **2024**, *14*, 1468. [CrossRef]

https://doi.org/10.3390/met16010060

metals MDPI

Article

Correlation Between Morphology and Crystal Structure of Electrolytically Produced Zinc Dendritic Particles

Nebojša D. Nikolić [1,*], Jelena D. Lović [1], Vesna M. Maksimović [2], Nikola S. Vuković [3], Nenad L. Ignjatović [4], Predrag M. Živković [5] and Sanja I. Stevanović [1]

[1] Department of Electrochemistry, Institute of Chemistry, Technology and Metallurgy, University of Belgrade, Njegoševa 12, 11000 Belgrade, Serbia; jelena.lovic@ihtm.bg.ac.rs (J.D.L.); sanjas@ihtm.bg.ac.rs (S.I.S.)
[2] Vinča Institute of Nuclear Science—National Institute of the Republic of Serbia, University of Belgrade, Mike Petrovića Alasa 12–14, 11000 Belgrade, Serbia; vesnam@vin.bg.ac.rs
[3] Institute for Technology of Nuclear and Other Mineral Raw Materials, Bulevar Franše d'Eperea 86, 11000 Belgrade, Serbia; n.vukovic@itnms.ac.rs
[4] Institute of Technical Sciences of the Serbian Academy of Science and Arts, Knez Mihailova 35/IV, 11000 Belgrade, Serbia; nenad.ignjatovic@itn.sanu.ac.rs
[5] Faculty of Technology and Metallurgy, University of Belgrade, Karnegijeva 4, 11000 Belgrade, Serbia; peca@tmf.bg.ac.rs
* Correspondence: nnikolic@ihtm.bg.ac.rs; Tel.: +381-11-337-0390

Abstract: The correlation between the morphology and crystal structure of zinc dendritic particles produced by electrolysis from the alkaline electrolyte has been established. Morphology and crystal structure of Zn particles electrodeposited by the potentiostatic regime of electrolysis at overpotentials inside (−100 and −160 mV) and outside (−220, −280, and −340 mV) the plateau of the limiting diffusion current density were characterized by scanning electron microscope (SEM) and by X-ray diffraction (XRD), respectively. The particle size distribution (PSD) was performed in order to determine the dependency of the size of dendritic particles on applied electrolysis overpotential. With increasing the overpotential of electrolysis, the shape of particles changed from irregular forms denoted as precursors of dendrites to various forms of dendrites, while the size of the particles simultaneously decreased. All types of Zn dendrites exhibited the strong (002) preferred orientation, while the precursors of dendrites exhibited (101)(002) preferred orientation. The development of strong (002) preferred orientation was explained and discussed by making an analogy with the electrolytic production of lead dendrites from the concentrated nitrate electrolyte. Although zinc and lead belong to different types of crystal lattice (Pb-face-centered cubic type and Zn-hexagonal close-packed type), they have a common characteristic that is manifested by the strong preferred orientation in the crystal plane with the lowest surface energy.

Keywords: zinc; electrolysis; dendrites; morphology; structure; SEM; XRD; PSD

1. Introduction

Zn is a metal that has attracted huge attention from both academic and technological communities for a long time due to its high availability, low price and toxicity, environmental acceptability, and high specific energy [1–3]. Also, Zn possesses high electrical conductivity and chemical stability in various aqueous electrolytes and has good electrochemical reversibility (−0.762 V relative to the standard hydrogen electrode (SHE)). Thanks to all these properties, it is widely used in the production of Zn-based rechargeable batteries, such as zinc–air, zinc–ion batteries, Zn-Ni batteries, and Zn-MnO_2 batteries [2,4,5].

This battery type belongs to a group of green and sustainable energy storage systems and simultaneously represents an excellent replacement for lithium–ion batteries. The morphology of Zn electrodes strongly affects the electrochemical performance of the Zn-based rechargeable batteries. For example, in Zn-MnO_2 batteries, the discharge capacity of 220 $mAhg^{-1}$ is delivered by porous Zn electrodes while only a discharge capacity of 130 $mAhg^{-1}$ was delivered by planar Zn electrodes [4]. This increase of 40% can largely be attributed to the porosity of the Zn anode and the subsequent increased surface area of the active metal. Namely, while zincate ions and ZnO products were found after the discharge at the surface of the planar anode, these products were not found in the porous anode, pointing out that the porosity prevents the detrimental buildup of zincate ions at the anodic surface that leads to passivation. Also, Zn is successfully used as a catalyst to electrochemically reduce CO_2 to CO [6].

The origin of the term "dendrite" is the Greek word *dendron*, which means a tree [3]. The classical definition of a dendrite is given by Wranglen [7], who defined a dendrite as a skeleton of a monocrystal that consists of a stalk and branches. The branches developed from the stalk are marked as primary (P) branches, while those developed from the P branches are marked as secondary (S) branches. If a stalk and branches are in the same plane, then a dendrite is marked as a 2D (two-dimensional) dendrite. If branches grow from a stalk in all directions, then the dendrite is marked as a 3D (three-dimensional) dendrite. The 2D dendrites with P branches, as well as those with P and S branches, are often marked as the fern-like dendrites [8]. The dendrites without branches with the sharp tips are known as the needle-like dendrites [8,9]. The dendrites belong to the group of powdered (dispersed and irregular) deposits and are formed during the diffusion control of electrodeposition [10].

Zinc dendrites can be obtained by the electrochemical deposition (ED) technique from alkaline [1,2,9–18], neutral [17,19,20], and acid [21–23] electrolytes. The other morphological forms that are formed by Zn electrolysis are spongy (mossy), boulders, and different grains of regular and irregular shapes [11,18,24,25]. All relevant parameters of electrolysis affecting the morphology of electrodeposited metals, such as electrolyte pH value, additives, current density, overpotential, temperature, etc., also affect Zn dendrite formation [2]. The electrolytic formation of Zn dendrites has both positive and negative sides. In secondary Zn–air batteries, the growth of Zn dendrites during the charging process can negatively affect the cycle life of the battery by penetrating the separator with sharp tips and by making a short circuit between the anode and the cathode [2,9,26]. In this way, the cycle life of a battery is shortened. Also, Zn dendrites as a type of powdered form, have a weak adhesion, and their spontaneous detachment from the electrode surface causes a loss of battery capacity. For that reason, the effort is also given on an inhibition of dendritic growth of zinc [2,27–29]. The most common way of inhibiting dendritic growth is the introduction of various compounds in zinc electrolytes, such as polymers, organic molecules, and metal ions [2,27,28]. The typical polymers added into electrolytes are polyethylenimine (PEI), Triton X-100, polyethylene glycol (PEG), polyvinyl alcohol (PVA), etc. [27]. Organic compounds, such as 4-aminomethyl cyclohexanecarboxylic acid (AMCA) [28] and many other compounds based on alcohols, aldehydes, organic acids, ammonium compounds, carbohydrates, and surfactants, are widely used for the inhibition of dendritic growth [27]. Finally, metal ions with more positive electrode potentials than zinc, such as bismuth ions (Bi^{3+}) and lead ions (Pb^{2+}), are also used for these purposes. The dendritic growth can be prevented by making a hybrid interface by which direct Zn–H_2O contact is fully avoided. It is the case in aqueous calcium ion batteries where an interphase layer with densely packed nanocrystals of $Ca_3(CO_3)_2(OH)_2 \cdot 1.5\ H_2O$ and ZnF_2 and amorphous organic species is demonstrated for a Zn metal anode with 1 M calcium trifluoromethyl sulfonate aqueous electrolyte [29]. The positive side of Zn dendrites is perceived through their use as catalysts for CO_2 conversion to CO [6]. Thanks to the very high specific surface area caused by branched morphology, Zn dendrites can show an order of magnitude higher activity and three-fold higher faradaic efficiency than bulk Zn [6].

These actual technologies are a cause of numerous and permanent investigations related to the formation and growth of dendrites [30–39]. Despite these investigations, there are still unknowns related to this topic. In the existing investigations, authors usually consider the formation and growth of dendrites, but without a detailed analysis of specific forms of dendrites that can be formed. Also, available information about the crystal structure of Zn dendrites is rare and unsystematic. For that reason, to complete existing knowledge, this study aimed to identify all possible shapes of Zn dendrites that can be obtained under various electrolysis conditions from the alkaline electrolyte and then to correlate the obtained morphology of Zn with their crystal structure.

2. Materials and Methods

Zinc was electrodeposited potentiostatically from the alkaline electrolyte at a temperature of 22.0 ± 0.50 °C in an electrochemical cell of a cylindrical shape. Electrodeposition was performed using a potentiostat/galvanostat BioLogic SP 200 (Seyssinet-Pariset, France) at overpotentials of −100, −160, −220, −280, and −340 mV. The composition of the electrolyte was 0.35 M ZnO in 6.0 M KOH. This composition was selected because these concentrations of ZnO and KOH, as well as their ratio, represent the typical values used in the Zn–air batteries. For the preparation of the electrolyte, double distilled water (Millipore, 18 M ohm cm resistivity; Merck KGaA, Darmstadt, Germany) and analytical grade reagents supplied by Sigma-Aldrich Company (Sigma-Aldrich, Burlington, MA, USA) were used.

The working electrode (WE) used for ED of Zn was of copper, while the counter (CE) and the reference (RefE) electrodes were of zinc. A Cu WE was cylindrical in shape (length: 2.0 cm; diameter: 0.080 cm) with an overall surface area of 0.50 cm^2 situated in the center of an electrochemical cell. The Zn CE was positioned about the wall of the electrochemical cell, while the tip of the reference electrode was positioned about 2 mm from the surface area of the Cu WE.

The cathodic polarization curve was recorded by the linear sweep voltammetry technique using the same WE, RefE, and CE electrodes as those used for Zn electrodeposition at selected overpotentials. A sweep rate of 1 mVs^{-1} was used during the recording of the polarization curve. All overpotentials are given relative to Zn reference electrode. For this electrodeposition system, Open Circuit Potential (OCP) was 912 ± 1.0 mV vs. Zn.

The preparation of Cu WE for the ED process takes place in several steps: (a) alkaline degreasing at a temperature of 70 °C; (b) rinsing in distilled water; (c) acid etching in 20% H_2SO_4 at 50 °C; and (d) rinsing in distilled water.

Characterization of Potentiostatically Produced Zn Dendritic Particles

Characterization of potentiostatically electrodeposited Zn particles was performed by techniques of scanning electron microscopy (SEM), X-ray diffraction (XRD), and particle size distribution (PSD). The following analyses were performed:

(a) *Morphology analysis*: Morphological analysis of potentiostatically electrodeposited Zn particles was performed by scanning electron microscope (SEM; model JEOL JSM-7001F JEOL Ltd., Tokyo, Japan). For morphological analysis, ED of Zn was performed with electrodeposited charges of 0.75, 1.5, 3.0, and 6.0 mAh, and Zn particles were characterized without their removal from the electrode surface after the finished ED process.

(b) *Structural analysis*: Structural characterization of potentiostatically electrodeposited Zn particles was performed by X-ray diffraction (XRD; Rigaku Ultima IV diffractometer, Rigaku Co., Ltd., Tokyo, Japan). Six reflections in a 2θ range between 30 and 75°

were recorded by this diffractometer with CuKα radiation to determine the preferred orientation of produced particles.

(c) *Analysis of the particle size*: The particle size distribution (PSD) of Zn particles was performed using the MALVERN Instruments MASTERSIZER 2000 device (MALVERN Instruments Ltd., Malvern, Worcestershire, UK).

For XRD and PSD analyses, electrodeposition of Zn was performed with an electrodeposited charge of 18.0 mAh. For these analyses, Zn particles were removed from the Cu WE after the completion of the electrolysis process. The electrolysis process was repeated multiple times to collect enough particles for the analysis. The same amounts of Zn particles were produced at all five overpotentials.

3. Results

3.1. Electrochemical Aspects of Electrolytic Production of Zn Dendritic Particles

The cathodic polarization curve for the ED process of Zn from an alkaline electrolyte containing 0.35 M ZnO in 6.0 M KOH is presented in Figure 1. The plateau of the limiting diffusion current density corresponds to an overpotential range between −90 and −180 mV, and the end of the plateau is marked by the beginning of the fast growth of the current density with the increase of overpotential. The electrolytic production of Zn in the form of dendrites commences within the plateau of the limiting diffusion current density, i.e., in the diffusion control of the electrodeposition, and continues to occur in a region of the fast growth of the current density after the inflection point at the polarization curve that denotes the end of the plateau.

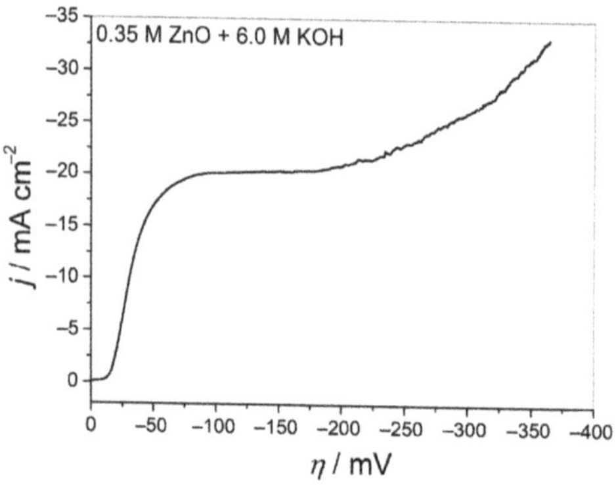

Figure 1. The cathodic polarization curve for electrodeposition of Zn from 0.35 M ZnO in 6.0 M KOH. Sweep rate: v = 1 mVs^{-1}.

The chronoamperometry curves recorded at overpotentials of −100 and −160 (the plateau of the limiting diffusion current density), −220, −280, and −340 mV (the zone of the fast increase of the current density with the increasing overpotential) are presented in Figure 2. The common feature of all chronoamperometric curves is a rise in the current density after a certain critical time, with a tendency to shorten this time with increasing the overpotential. The beginning of the growth of the current density coincides with an initiation of dendritic growth [40]. Namely, a growth of dendrites causes a considerable increase of the cathode surface during the electrolysis process [41]. In a potentiostatic regime of electrolysis, since the real current density stays constant during the electrolysis duration, while the cathode surface area increases with time, the final result is an increase of the current density with the electrolysis time in the chronoamperometry curves. According to recorded chronoamperometry curves, the growth of dendrites commences approximately after 810 s at −100 mV, 550 s at −160 mV, 90 s at −220 mV, 45 s at −280 mV, and 19 s at −340 mV.

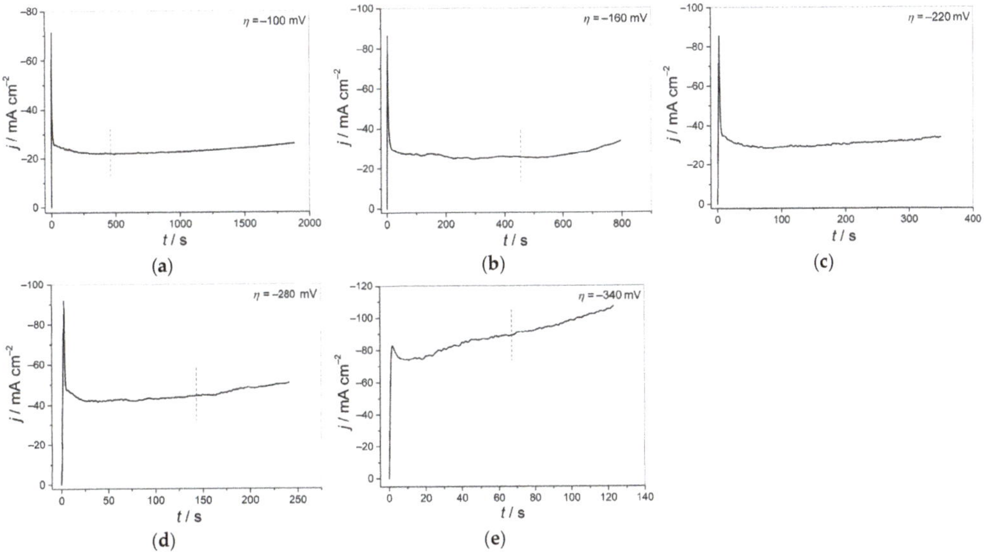

Figure 2. The chronoamperometric dependencies recorded from 0.35 M ZnO in 6.0 M KOH at overpotentials, η, of (**a**) −100 mV; (**b**) −160 mV; (**c**) −220 mV; (**d**) −280 mV; and (**e**) −340 mV. The vertical dashed lines in (**a**,**b**) correspond to an electrodeposited charge of 1.5 mAh, while those given in (**d**,**e**) correspond to an electrodeposited charge of 0.75 mAh.

3.2. Morphological Analysis of Zn Irregular Forms

The morphology of the zinc deposit obtained at an overpotential of −100 mV, belonging to the very beginning of the plateau of the limiting diffusion current density with an electrodeposited charge of 1.5 mAh, is shown in Figure 3. This electrodeposited charge is denoted by a vertical dashed line in Figure 2a, from which can be seen that the time that corresponded to it was below the critical value for initiation of dendritic growth. As supposed, dendrites were not formed under these electrolysis conditions. Instead of dendrites, grains of both irregular and regular shapes with well-defined crystal planes were formed (Figure 3 and Figure S1 in Supplementary Materials). Some of the grains had regular hexagonal shapes.

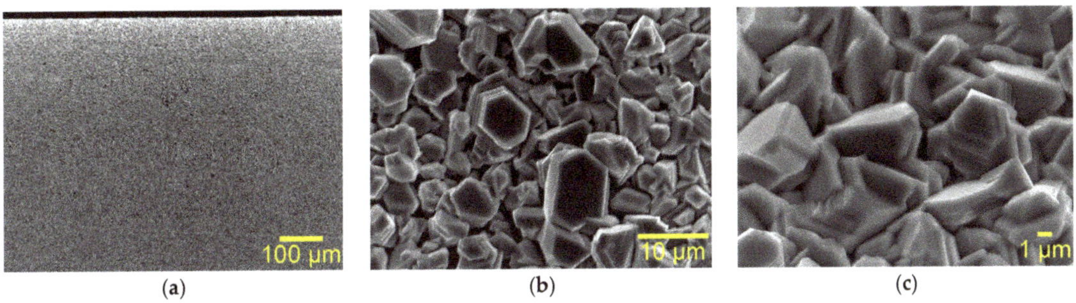

Figure 3. Morphologies of Zn deposit obtained from 0.35 M ZnO in 6.0 M KOH at an overpotential of −100 mV with an electrodeposited charge of 1.5 mAh: (**a**) top view and (**b**,**c**) details from figure under (**a**) recorded at the higher magnifications. The time of electrodeposition: 460 s.

The morphology of the Zn deposit electrodeposited at the same overpotential, but with an electrodeposited charge of 6.0 mAh, is shown in Figure 4. This electrodeposited charge corresponds to an electrolysis time of 1880 s, which was above the critical value for initiation of dendritic growth (Figure 2a). Electrodeposition with this electrodeposited charge led to a formation of irregular forms with dendritic character (Figure 4a,b). Some of the typical forms of dendritic character are also given in the Supplementary Materials (see Figure S2). These forms do not follow the classical definition of a dendrite given by Wranglen [7], and they can be denoted as precursors of dendrites. Aside from them, both irregular and regular grains were also formed (Figure 4c).

(**a**) (**b**) (**c**)

Figure 4. Morphologies of Zn deposit obtained from 0.35 M ZnO in 6.0 M KOH at an overpotential of −100 mV with an electrodeposited charge of 6.0 mAh: (**a**) top view and (**b**,**c**) details from figure under (**a**) recorded at the higher magnifications. The time of electrodeposition: 1880 s.

Figure 5 shows the morphology of the Zn deposit obtained at an overpotential of −160 mV with an electrodeposited charge of 1.5 mAh. The time corresponding to this electrodeposited charge is denoted by the vertical dashed line in Figure 2b and was below the critical value for an initiation of dendritic growth. Similar to the electrodeposition process at −100 mV, grains of both irregular and regular shapes with well-defined crystal planes were formed with this electrodeposited charge (Figure 5a–c). Some of the grains also have regular hexagonal shapes (Figure 5b). Electrodeposition with this electrodeposited charge led to the appearance of the irregular forms of dendritic character, too (Figure 5c). More grains of regular and irregular shapes are shown in Figure S3 in Supplementary Materials.

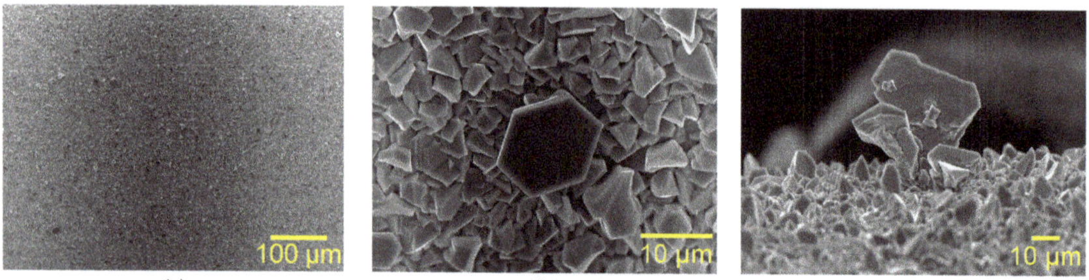

(**a**) (**b**) (**c**)

Figure 5. Morphologies of Zn deposit obtained from 0.35 M ZnO in 6.0 M KOH at an overpotential of −160 mV with an electrodeposited charge of 1.5 mAh: (**a**) top view and (**b**,**c**) details from figure under (**a**) recorded at the higher magnifications. The time of electrodeposition: 452 s.

The morphology of the Zn deposit electrodeposited at the same overpotential, but with an electrodeposited charge of 3.0 mAh, is shown in Figure 6. This electrodeposited charge corresponds to an electrolysis time of 800 s, which was above the critical value for initiation of dendritic growth (Figure 2b). After 800 s the current density was 30.5% larger than that after 550 s, which corresponds to the initiation of dendritic growth. Both irregular (Figure 6a,b) and regular (Figure 6c) dendrites oriented by the tips towards the bulk of the electrolyte were formed with this electrodeposited charge. It is worth noting that regular dendrites had compact, massive shapes. Some forms of irregular and regular dendrites are also shown in Figure S4 in Supplementary Materials.

Figure 6. Morphologies of Zn deposit obtained from 0.35 M ZnO in 6.0 M KOH at an overpotential of −160 mV with an electrodeposited charge of 3.0 mAh: (**a**) top view and (**b**,**c**) details from figure under (**a**) recorded at the higher magnifications. The time of electrodeposition: 800 s.

The mixture of branchy 2D (two-dimensional) and 3D (three-dimensional) dendrites was formed at an overpotential of −220 mV with an electrodeposited charge of 1.5 mAh (Figures 7a and S5a). Formation of the dendrites with this electrodeposited charge was expected since the electrolysis time that corresponded to it was above the critical value for an initiation of dendritic growth (Figure 2c). Both 2D (Figure 7a,b) and 3D (Figure 7a,c) dendrites were oriented by the tips towards the bulk of the electrolyte, causing a considerable increase of the electrode surface during the duration of the electrolysis process.

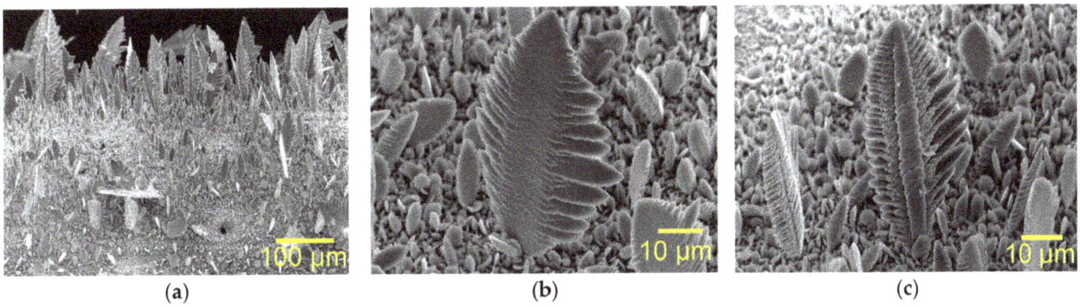

Figure 7. Morphologies of Zn deposit obtained from 0.35 M ZnO in 6.0 M KOH at an overpotential of −220 mV with an electrodeposited charge of 1.5 mAh: (**a**) top view and (**b**,**c**) the types of dendrites. The time of electrodeposition: 350 s.

The similar shapes of the dendrites to those obtained at an overpotential of −220 mV were also formed at overpotentials of −280 mV (Figure 8) and −340 mV (Figure 9) with the same electrodeposited charge. The mixture of branched 2D (Figures 8a,b, 9a,b and S5b,c) and 3D (Figures 8a,c, 9a,c and S5b,c) dendrites oriented by the tips towards the bulk of the electrolyte was formed at −280 and −340 mV. As expected, due to the larger applied overpotential of the electrodeposition, a larger number of the dendrites was formed at overpotentials of −280 and −340 mV than at −220 mV.

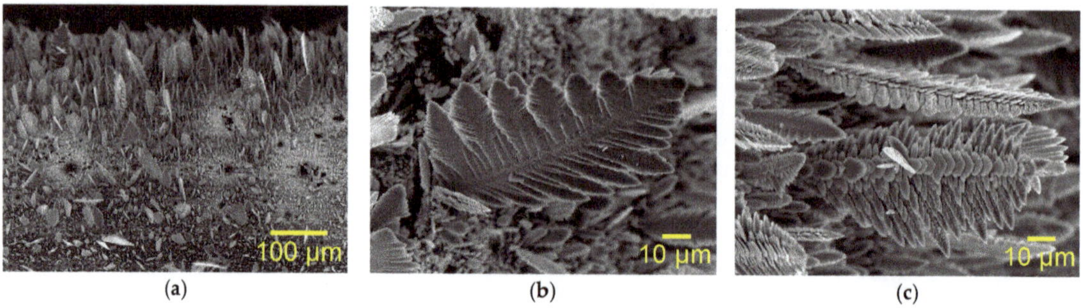

Figure 8. Morphologies of Zn deposit obtained from 0.35 M ZnO in 6.0 M KOH at an overpotential of −280 mV with an electrodeposited charge of 1.5 mAh: (**a**) top view and (**b**,**c**) the types of dendrites. The time of electrodeposition: 242 s.

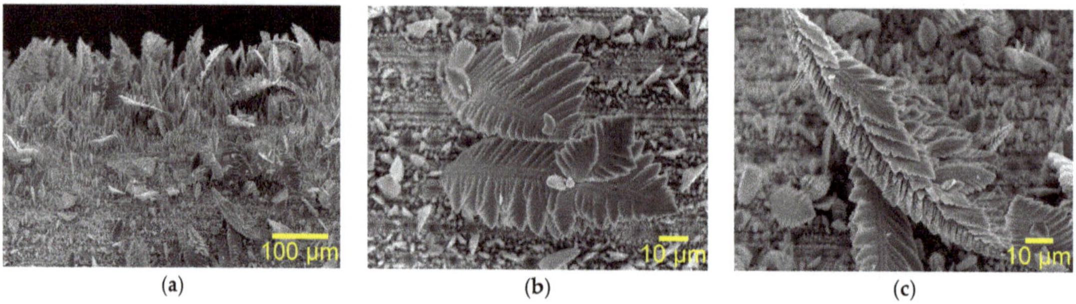

Figure 9. Morphologies of Zn deposit obtained from 0.35 M ZnO in 6.0 M KOH at an overpotential of −340 mV with an electrodeposited charge of 1.5 mAh: (**a**) top view and (**b**,**c**) the types of dendrites. The time of electrodeposition: 123 s.

3.3. The Crystal Structure of Zn Dendritic Particles

The XRD (X-ray diffraction) patterns of Zn dendritic particles electrodeposited at overpotentials of −100, −160, −220, −280, and −340 mV are presented in Figure 10a–e, respectively. The Joint Committee on Powder Diffraction Standard (JCPDS number 00-04-0831) for Zn powder established for spherical distributed grains is shown in Figure 10f. The sharp diffraction peaks observed at 2θ angles of 36.30°, 39.0°, 43.23°, 54.33°, 70.05°, and 70.66° corresponded to (002), (100), (101), (102), (103), and (110) crystal planes of Zn, simultaneously confirming the electrocrystallization of Zn in the hexagonal closed pack type of crystal lattice [42]. It is worth noting that no other diffraction peaks were observed, pointing out a high purity of synthesized Zn particles.

At first glance, it can be seen that Zn crystallites in the precursors of dendrites formed at −100 mV were oriented in larger extent in the (101) and (002) crystal planes (Figure 10a), while those in the dendrites produced at −160, −220, −280, and −340 mV were predominately oriented in the (002) crystal plane (Figure 10b–e). Simultaneously, all diffractograms strongly differed from the JCPDS standard for Zn. This difference represented a good indicator for the existence of preferred orientation in potentiostatically produced Zn particles. To establish it, the preferred orientation of Zn particles was estimated by applying a methodology based on a determination of "Texture Coefficients" (TCs(hkl)) and "Relative Texture Coefficients" (RTCs(hkl)) [8,43]. The values of TCs(hkl) coefficients larger than 1 point out the existence of the preferred orientation, while the values of RTCs(hkl) coefficients are dependent on the number of recorded diffraction peaks [8,40,43]. In our case, six main Zn diffraction peaks were recorded, and then the values of RTCs(hkl) coefficients larger than 16.7% pointed out the existence of the preferred orientation. The basis of this methodology, together with relative intensities of the diffraction peaks, is given in Supplementary Materials, while the values of TCs(hkl) and RTCs(hkl) coefficients are given in Tables 1 and S1 for the particles produced at overpotentials belonging to the plateau of the limiting diffusion current density (−100 mV and −160 mV) and Tables 2 and S1 for the particles obtained at overpotentials outside the plateau of the limiting diffusion current density (−220 mV, −280 mV, and −340 mV).

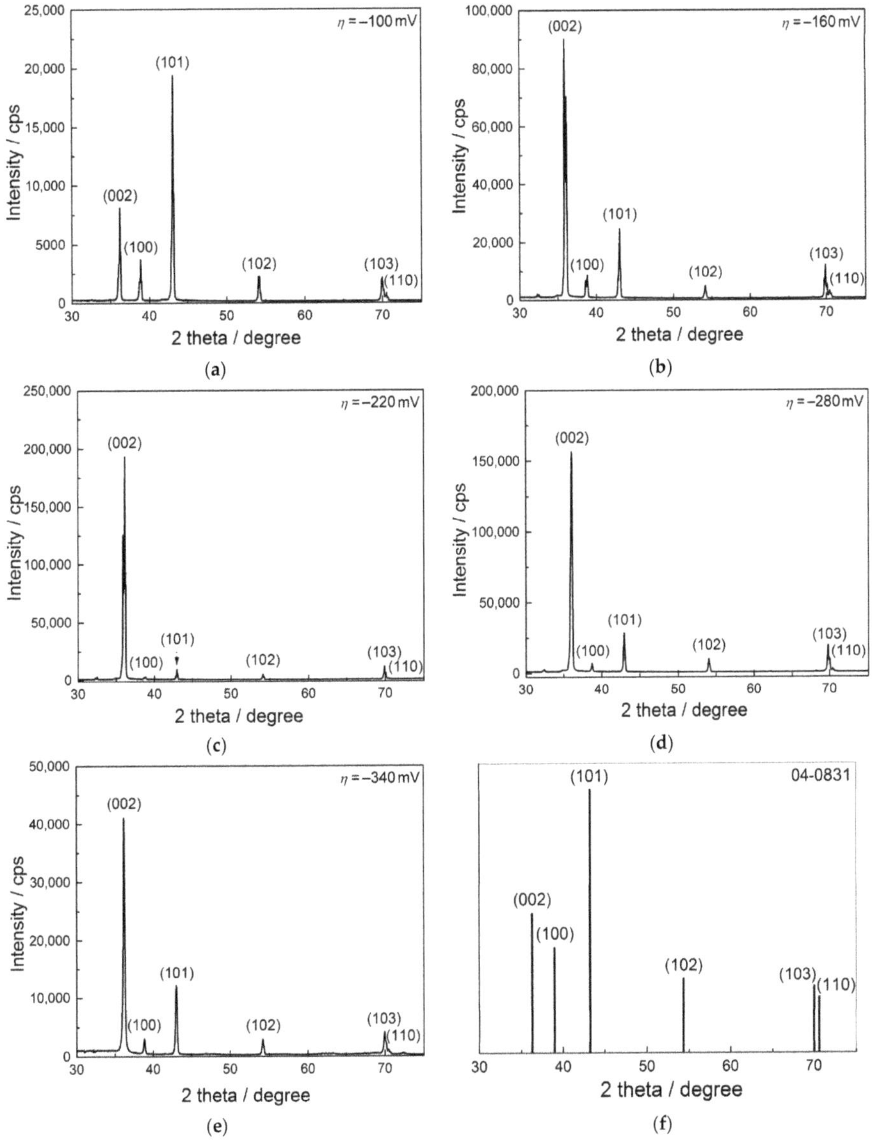

Figure 10. X-ray diffraction (XRD) patterns obtained for Zn particles electrodeposited from 0.35 M ZnO in 6.0 M KOH at overpotentials, η, of (**a**) −100 mV; (**b**) −160 mV; (**c**) −220 mV; (**d**) −280 mV; (**e**) −340 mV; and (**f**) Zn standard (04-0831).

Table 1. The values of *TCs(hkl)* and *RTCs(hkl)* coefficients for Zn particles produced at overpotentials, η, of −100 and −160 mV.

Crystal Plane	η = −100 mV		η = −160 mV	
	TCs(hkl)	*RTCs(hkl)*/%	*TCs(hkl)*	*RTCs(hkl)*/%
(002)	1.1	23.7	3.1	55.5
(100)	0.67	14.4	0.39	7.0
(101)	1.4	30.1	0.45	8.0
(102)	0.59	12.7	0.31	5.5
(103)	0.62	13.3	0.87	15.6
(110)	0.27	5.8	0.47	8.4

Table 2. The values of *TCs(hkl)* and *RTCs(hkl)* coefficients for Zn particles produced at overpotentials, η, of −220, −280, and −340 mV.

Crystal Plane	η = −220 mV		η = −280 mV		η = −340 mV	
	TCs(hkl)	*RTCs(hkl)*/%	*TCs(hkl)*	*RTCs(hkl)*/%	*TCs(hkl)*	*RTCs(hkl)*/%
(002)	4.3	77.4	3.5	62.9	3.2	60.5
(100)	0.087	1.6	0.21	3.8	0.26	4.9
(101)	0.11	2.0	0.34	6.1	0.50	9.5
(102)	0.20	3.6	0.42	7.6	0.44	8.3
(103)	0.54	9.7	0.93	16.7	0.67	12.7
(110)	0.32	5.7	0.16	2.9	0.22	4.1

Analysis of data collected in Tables 1 and 2 pointed out that precursors of dendrites produced at −100 mV showed (101)(002) preferred orientation, while dendrites of various shapes produced at −160, −220, −280, and −340 mV exhibited the strong (002) preferred orientation.

3.4. The Dependence of Size of Zn Dendritic Particles on Applied Overpotential of the Electrodeposition

The effect of electrodeposition overpotential on the size of the dendritic particles is examined by particle size distribution (PSD) analysis of the particles obtained at various overpotentials with the same electrodeposited charge. The PSD dependencies obtained for the particles synthesized at the given overpotentials are shown in Figure 11, from which it can be seen that the particle size decreased with increasing the overpotential of electrodeposition. The uniform distribution was obtained for the dendrites synthesized at overpotentials of −220, −280, and −340 mV, while non-uniform distribution was obtained for the particles electrodeposited at overpotentials of −100 and −160 mV.

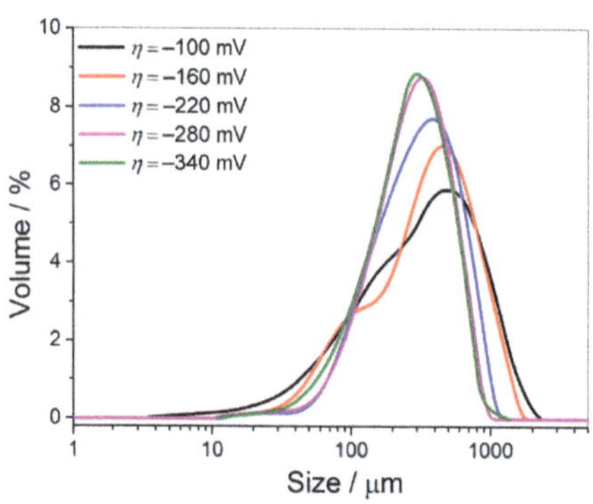

Figure 11. Particle size distribution (PSD) of Zn electrodeposited at overpotentials of −100, −160, −220, −280, and −340 mV.

4. Discussion

The dendrites were the main morphological shapes formed by Zn electrolysis from the alkaline electrolyte. The shape of the dendrites strongly depended on the applied overpotential of the electrolysis and changed from irregular and compact, massive ones formed at the overpotential within the plateau of the limiting diffusion current density to the branchy 2D and 3D dendrites formed at overpotentials outside the limiting diffusion current density plateau. The particles of irregular shape denoted as precursors of dendrites were formed at the very beginning of the plateau of the limiting diffusion current density. The dendrites are formed in the diffusion control of electrodeposition, and various models, such as the "pyramid" growth model based on spiral dislocation growth controlled by bulk diffusion [44], diffusion-limited aggregation and phase–field models [45,46], a quantitative phase–field model based on Marcus kinetics theory [47], etc., have been proposed to explain the formation and growth of Zn dendrites.

The general theory of disperse deposits formation [10,11,48] based on the Diggle, Despić, and Bockris theory [44] and later upgraded by investigations of Popov and Nikolić [48]. It is one of a well-established theory that successfully explains the formation and growth of dendrites in general. According to this theory, a dendrite originates from a protrusion or irregularity formed in an initial phase of electrolysis, around which the tip of the spherical diffusion layer is formed. The current density at the tip of a protrusion is larger than on the flat part of the electrode surface, causing the formation of a dendrite as the final morphological form in the growth process. The tip of protrusion grows under the activation control, while simultaneously, the electrolysis process on the flat part of the electrode surface occurs predominately under the full diffusion control [10,11,44,48]. Regarding the polarization curve shown in Figure 1, it is necessary to point out that the electrolysis process remains diffusion-controlled after the end of the limiting diffusion current density plateau, and that the fast increase of the current density with the further increase of overpotential is a result of the increase of the real electrode surface caused by the fast growth of dendrites.

The electrodeposited charge, i.e., the electrodeposition time, does not affect the morphology but only the size of dendrites. Figure 12 shows the morphologies of dendrites obtained at overpotentials of −280 mV (Figure 12a–c) and −340 mV (Figure 12d–f) with an electrodeposited charge of 0.75 mAh. This electrodeposited charge is denoted by a vertical dashed line in Figure 2d,e and was above the critical times for an initiation of dendritic growth at both overpotentials. Comparing the shapes of dendrites obtained at the same overpotentials with different electrodeposited charges (Figures 8 and 12a–c; −280 mV) and (Figures 9 and 12d–f; −340 mV), it can be seen that they are the same in shape but different in size. It can be explained by the fact that the real current density always remains constant during electrodeposition in a potentiostatic regime [41] as follows: the dendritic growth is initiated at some overpotential belonging to the plateau of the limiting diffusion current density, and this overpotential is denoted as critical overpotential for an initiation of dendritic growth [10,11]. However, the growth of dendrites does not commence instantly at a given overpotential but after a certain time corresponding to a time necessary for an initiation of dendritic growth. After this time, due to the growth of the dendrites, the cathode surface starts to increase, and it is manifested by the rise of the current density in chronoamperometry curves (Figure 2). This rise in the current density keeps the real current density constant during the whole electrodeposition process, keeping the shape of the dendrites simultaneously unchanged. The effect of the electrodeposited charge is shown for overpotentials of −280 and −340 mV, but it is also valid for other overpotentials.

(a) (b) (c) (d) (e) (f)

Figure 12. Morphologies of Zn deposits obtained from 0.35 M ZnO in 6.0 M KOH at overpotentials of (**a–c**) −280 mV and (**d–f**) −340 mV with an electrodeposited charge of 0.75 mAh. The time of electrodeposition: (**a–c**) −280 mV: 143 s and (**d–f**) −340 mV: 67 s.

Aside from the strong effect of the overpotential of electrodeposition on the shape of the dendritic particles, the applied overpotentials also affected their size (Figure 11). The dependence of the particle size on the overpotential of the electrodeposition can be explained as follows: The dependency of the nucleation rate, J, on the overpotential of the electrodeposition, η, is given by Equation (1) [10].

$$J = K_1 \exp\left(-\frac{K_2}{\eta^2}\right) \quad (1)$$

In Equation (1), K_1 and K_2 are constants independent of overpotential. Analysis of Equation (1) shows that the nucleation rate increases with the increase of the electrodeposition overpotential. It means that a larger number of nuclei, and hence, protrusions of which the dendrites are formed, arise at higher rather than at lower overpotentials. This will cause the formation of a larger number of dendrites in the growth process at higher than at lower overpotentials. Since the same electrodeposited charge is spent for the dendrite formation at all overpotentials, it is clear that it will form a larger number of smaller dendrites at higher than at lower overpotentials. The obtained PSD dependencies just confirmed this, since it is very clear from Figure 11 that the decrease in dendrite size occurs together with the increase in overpotential of the electrodeposition.

As already mentioned, the precursors of dendrites showed (101)(002) preferred orientation, while the dendrites showed a strong (002) preferred orientation. To explain the strong (002) preferred orientation, an analogy with lead electrolysis processes from the concentrated nitrate electrolyte was made. Pb and Zn belong to the same group of metals from an electrochemical point of view but to different types of crystal lattices. Regarding electrochemical parameters, they belong to a group of so-called normal metals, characterized by high exchange current density and overpotential for hydrogen evolution reaction and by low melting point values [49]. These parameters enable us to reach the diffusion control at relatively low overpotentials and to form powdered (disperse; irregular) deposits in a whole range of overpotentials [10]. On the other hand, Pb belongs to the face-centered cubic type, while Zn belongs to the hexagonal close-packed type of crystal lattice. The irregular and the 2D dendritic forms having the strong (111) preferred orientation were formed during Pb electrolysis [8]. In a face-centered cubic lattice, the (111) crystal plane is a plane with the lowest surface energy. This plane survived a growth process during the duration of electrolysis as a slow-growing plane, constructing an interior, i.e., a body of dendrites. The other crystal planes, such as (220), (200), and (311), are fast-growing planes that disappear during the growth process. The tips, edges, and corners of dendrites were constructed from these crystal planes [50].

Similar to the face-centered cubic type of crystal lattice, each crystal plane in the hexagonal close-packed type of a crystal lattice also has a different arrangement of atoms [51]. In this type of crystal lattice, the (002) is a basal crystal plane with Zn atoms that are orderly arranged in a regular hexagon structure with a flat surface, where each atom of zinc is connected with nine surrounding atoms [51–55]. This arrangement of Zn atoms means that the (002) crystal plane has lower surface energy than all other crystal planes. Then, it is clear that this crystal plane is equivalent to the (111) crystal plane in the face-centered cubic crystal lattice of Pb. Hence, the (002) crystal plane represents a slow-growing plane, and as such, it remains during Zn electrolysis from the alkaline electrolyte, constructing the interior, i.e., body or stalk of Zn dendrites. It is confirmed by morphological analysis of the Zn dendrite electrodeposited at an overpotential of −160 mV (Figure 13a). The regular hexagons formed inside the dendrite are identified, as seen in Figure 13b. Furthermore, the stacked hexagonal structures can easily be identified from Figure 13c, which presents the branches of the 3D dendrite. The other crystal planes, such as (100), (101), (102), (103), and (110), with higher surface energies and Zn atoms arranged in a more irregular wavy structure than the (002) crystal plane, can be denoted as the fast-growing crystal planes that disappear during the growth process, constructing tips, edges, and corners of the Zn dendrites. Simultaneously, it is clear that regular hexagons formed at overpotentials of −100 and −160 mV (Figures 3b, 4c and 5b) correspond to the (002) crystal plane.

(a) (b) (c)

Figure 13. (**a**) Zn dendrite obtained from 0.35 M ZnO in 6.0 M KOH at an overpotential of −160 mV with an electrodeposited charge of 1.5 mAh; (**b**) details from the dendrite shown in figure under (**a**); and (**c**) the branches of 3D dendrites electrodeposited at an overpotential of −220 mV with an electrodeposited charge of 1.5 mAh (a detail from the dendrite shown in Figure 7c).

The strong correlation between the morphology of dendrites and the preferred orientation in the crystal plane with the lowest surface energy can be confirmed by the analysis of tin electrodeposition processes. Sn belongs to the same group of metals from an electrochemical point of view as Pb and Zn (the group of so-called normal metals) but to a different type of crystal lattice, i.e., to the body-centered tetragonal type. The electrolytically produced dendrites of Sn from the alkaline electrolyte exhibited the strong (220)(440) preferred orientation, which are the crystal planes with the lowest surface energy for this type of crystal lattice [40]. Hence, the electrochemical parameters prevail over those belonging to the determined type of crystal lattice.

The high purity of the produced dendritic particles can also be attributed to the predominant orientation of Zn crystallites in the (002) crystal plane. As the crystal plane with the lowest surface energy, the (002) crystal plane has more stable thermodynamic properties than all other crystal planes, making this plane electrochemically less active than the other crystal planes [51,55]. This causes the Zn particles with this predominant preferred orientation to have better corrosion resistance than those with other preferred orientations.

5. Conclusions

Morphological and structural analysis of zinc dendritic particles was performed to establish a correlation between these two characteristics of particles. The dendrites of Zn were produced potentiostatically from the alkaline electrolyte at overpotential, which belonged to the plateau of the limiting diffusion current density (−160 mV), as well as at those outside the plateau in the zone of the fast growth of the current density with increasing the overpotential (−220 mV, −280 mV, and −340 mV). The particles of irregular shape with dendritic character denoted as precursors of dendrites were formed at the very beginning of the plateau of the limiting diffusion current density (−100 mV). The concentrations of 0.35 M ZnO and 6.0 M KOH were in the range of values typically used in the secondary Zn–air batteries. The obtained particles were characterized by scanning electron microscope (SEM; morphology) and X-ray diffraction (XRD; crystal structure), while the dependency of the particle size on the electrolysis overpotential was estimated by the particle size distribution (PSD) analysis.

On the basis of the results of performed analyses, it can be concluded that the shape of dendrites strongly depended on the applied overpotential of electrolysis. The two types of dendrites were identified by electrolysis from the alkaline electrolyte: (a) irregular and compact massive dendrites formed at the overpotential inside the plateau of the limiting diffusion current density, and (b) very branchy, both 2D (two-dimensional) and 3D (three-dimensional) dendrites formed at overpotentials outside the plateau of the limiting diffusion current density. Irregular and regular grains with well-defined crystal planes, including those hexagonal shapes, were concurrently formed beside dendrites at the overpotentials that belonged to the limiting diffusion current density plateau. The electrodeposited charge had only an influence on the size of dendrites, not on their shape. Simultaneously, the size of Zn dendritic particles decreased, while their uniformity increased with the increase of the overpotential of the electrolysis.

All types of dendrites showed a strong (002) preferred orientation, while precursors of the dendrites had (101)(002) preferred orientation. The strong (002) preferred orientation is explained by making an analogy with processes of lead electrolysis from the concentrated electrolyte. Pb and Zn belong to the same group of metals from the electrochemical point of view (so-called normal metals) and to various types of crystal lattice (Pb-face-centered cubic type and Zn-hexagonal close-packed type). The (002) crystal plane, as a plane with the lowest surface energy, builds the body or stalk of Zn dendrites, while tips, edges, and corners of the Zn dendrites are constructed from Zn crystallites oriented in other crystal planes.

Supplementary Materials: The following supporting information can be downloaded at: https://www.mdpi.com/article/10.3390/met14121468/s1. Figure S1: Zn grains of various shape obtained by the electrodeposition from 0.35 M ZnO in 6.0 M KOH at an overpotential of −100 mV with an electrodeposited charge of 1.5 mAh; Figure S2: Various forms of dendritic character obtained by the electrodeposition from 0.35 M ZnO in 6.0 M KOH at an overpotential of −100 mV with an electrodeposited charge of 6.0 mAh; Figure S3: Zn grains of irregular and regular shapes with well-defined crystal planes obtained by the electrodeposition from 0.35 M ZnO in 6.0 M KOH at an overpotential of −160 mV with an electrodeposited charge of 1.5 mAh; Figure S4: Zn dendrites obtained by the electrodeposition from 0.35 M ZnO in 6.0 M KOH at an overpotential of −160 mV with an electrodeposited charge of 3.0 mAh: (a) irregular shape, and (b) regular shape; Figure S5: The mixture of branchy 2D (two-dimensional) and 3D (three-dimensional) dendrites obtained by the electrodeposition from 0.35 M ZnO in 6.0 M KOH with an electrodeposited charge of 1.5 mAh at overpotentials of: (a) −220 mV, (b) −280 mV, and (c) −340 mV; Table S1: Texture calculations for Zn dendritic particles produced potentiostatically at overpotentials of −100, −160, −220, −280 and −340 mV. *R*—Intensity of the diffraction peak; *TC*—Texture Coefficient; *RTC*—Relative Texture Coefficient; s—Zn standard.

Author Contributions: Conceptualization, N.D.N.; methodology, N.D.N.; validation, N.D.N., J.D.L., N.L.I., S.I.S., P.M.Ž. and V.M.M.; investigation, N.D.N., J.D.L., N.L.I., S.I.S., N.S.V. and V.M.M.; resources, N.D.N., N.L.I., N.S.V. and V.M.M.; writing—original draft preparation, N.D.N. and P.M.Ž.; writing—review and editing, N.D.N., J.D.L., N.L.I., S.I.S. and V.M.M.; visualization, N.D.N.; supervision, N.D.N.; project administration, N.D.N.; funding acquisition, N.D.N. All authors have read and agreed to the published version of the manuscript.

Funding: This research was funded by the Ministry of Science, Technological Development, and Innovation of the Republic of Serbia (Contract No. 451-03-66/2024-03/200026) and the Science Fund of RS (Grant No. AdCatFC: 7739802).

Data Availability Statement: The raw data supporting the conclusions of this article will be made available by the authors on request.

Conflicts of Interest: The authors declare no conflicts of interest.

References

1. Pei, P.; Wang, K.; Ma, Z. Technologies for extending zinc–air battery's cyclelife: A review. *Appl. Energy* **2014**, *128*, 315–324. [CrossRef]
2. Zuo, Y.; Wang, K.; Pei, P.; Wei, M.; Liu, X.; Xiao, Y.; Zhang, P. Zinc dendrite growth and inhibition strategies. *Mater. Today Energy* **2021**, *20*, 100692. [CrossRef]
3. Zheng, X.; Ahmad, T.; Chen, W. Challenges and strategies on Zn electrodeposition for stable Zn-ion batteries. *Energy Storage Mater.* **2021**, *39*, 365–394. [CrossRef]
4. Minakshi, M.; Appadoo, D.; Martin, D.E. The Anodic Behavior of Planar and Porous Zinc Electrodes in Alkaline Electrolyte. *Electrochem. Solid-State Lett.* **2010**, *13*, A77–A80. [CrossRef]
5. Minakshi, M.; Ionescu, M. Anodic behavior of zinc in Zn-MnO_2 battery using ERDA technique. *Int. J. Hydrogen Energy* **2010**, *35*, 7618–7622. [CrossRef]
6. Rosen, J.; Hutchings, G.S.; Lu, Q.; Forest, R.V.; Moore, A.; Jiao, F. Electrodeposited Zn Dendrites with Enhanced CO Selectivity for Electrocatalytic CO_2 Reduction. *ACS Catal.* **2015**, *5*, 4586–4591. [CrossRef]
7. Wranglen, G. Dendrites and Growth Layers in the Electrocrystallization of Metals. *Electrochim. Acta* **1960**, *2*, 130–146. [CrossRef]
8. Nikolić, N.D.; Maksimović, V.M.; Avramović, L. Correlation of Morphology and Crystal Structure of Metal Powders Produced by Electrolysis Processes. *Metals* **2021**, *11*, 859. [CrossRef]
9. Fu, J.; Cano, Z.P.; Park, M.G.; Yu, A.; Fowler, M.; Chen, Z. Electrically Rechargeable Zinc–Air Batteries: Progress, Challenges, and Perspectives. *Adv. Mater.* **2017**, *29*, 1604685. [CrossRef] [PubMed]
10. Popov, K.I.; Djokić, S.S.; Nikolić, N.D.; Jović, V.D. *Morphology of Electrochemically and Chemically Deposited Metals*; Springer: New York, NY, USA, 2016; pp. 1–368.
11. Nikolić, N.D.; Živković, P.M.; Lović, J.D.; Branković, G. Application of the general theory of disperse deposits formation in an investigation of mechanism of zinc electrodeposition from the alkaline electrolytes. *J. Electroanal. Chem.* **2017**, *785*, 65–74. [CrossRef]
12. Nikitin, V.S.; Ostanina, T.N.; Rudoi, V.M.; Kuloshvili, T.S.; Darintseva, A.B. Features of Hydrogen Evolution during Electrodeposition of Loose Deposits of Copper, Nickel and Zinc. *J. Electroanal. Chem.* **2020**, *870*, 114230. [CrossRef]
13. Ostanina, T.N.; Rudoi, V.M.; Nikitin, V.S.; Darintseva, A.B.; Demakov, S.L. Change in the physical characteristics of the dendritic zinc deposits in the stationary and pulsating electrolysis. *J. Electroanal. Chem.* **2017**, *784*, 13–24. [CrossRef]

14. Ostanina, T.N.; Rudoi, V.M.; Patrushev, A.V.; Darintseva, A.B.; Farlenkov, A.S. Modelling the Dynamic Growth of Copper and Zinc Dendritic Deposits under the Galvanostatic Electrolysis Conditions. *J. Electroanal. Chem.* **2015**, *750*, 9–18. [CrossRef]
15. Wang, R.Y.; Kirk, D.W.; Zhang, G.X. Effects of deposition conditions on the morphology of zinc deposits from alkaline zincate solutions. *J. Electrochem. Soc.* **2006**, *153*, C357–C364. [CrossRef]
16. Banik, S.J.; Akolkar, R. Suppressing Dendritic Growth during Alkaline Zinc Electrodeposition using Polyethylenimine Additive. *Electrochim. Acta* **2015**, *179*, 475–481. [CrossRef]
17. Li, C.; Wang, L.; Zhang, J.; Zhang, D.; Du, J.; Yao, Y.; Hong, G. Roadmap on the protective strategies of zinc anodes in aqueous electrolyte. *Energy Storage Mater.* **2022**, *44*, 104–135. [CrossRef]
18. Jabbari, V.; Foroozan, T.; Shahbazian-Yassar, R. Dendritic Zn Deposition in Zinc-Metal Batteries and Mitigation Strategies. *Adv. Energy Sustain. Res.* **2021**, *2*, 2000082. [CrossRef]
19. Wang, C.; Li, J.; Zhou, Z.; Pan, Y.; Yu, Z.; Pei, Z.; Zhao, S.; Wei, L.; Chen, Y. Rechargeable zinc-air batteries with neutral electrolytes: Recent advances, challenges, and prospects. *Energy Chem.* **2021**, *3*, 100055. [CrossRef]
20. Thomas Goh, F.W.; Liu, Z.; Andy Hor, T.S.; Zhang, J.; Ge, X.; Zong, Y.; Yu, A.; Khoo, W. A Near-Neutral Chloride Electrolyte for Electrically Rechargeable Zinc-Air Batteries. *J. Electrochem. Soc.* **2014**, *161*, A2080. [CrossRef]
21. Bengoa, L.N.; Bruno, S.; Lazzarino, H.A.; Seré, P.R.; Egli, W.A. Study of dendritic growth of zinc crystals on the edges of steel sheet. *J. Appl. Electrochem.* **2014**, *44*, 1261–1269. [CrossRef]
22. Bengoa, L.N.; Seré, P.R.; Conconi, M.S.; Egli, W.A. Morphology and Texture of Zinc Deposits Formed at the Edge of a Rotating Washer Electrode. *J. Mater. Eng. Perform.* **2016**, *25*, 2936–2942. [CrossRef]
23. Bengoa, L.N.; Pary, P.; Seré, P.R.; Conconi, M.S.; Egli, W.A. Dendritic Zinc Growth in Acid Electrolyte: Effect of the pH. *J. Mater. Eng. Perform.* **2018**, *27*, 1103–1108. [CrossRef]
24. Desai, D.; Wei, X.; Steingart, D.A.; Banerjee, S. Electrodeposition of preferentially oriented zinc for flow-assisted alkaline batteries. *J. Power Sources* **2014**, *256*, 145–152. [CrossRef]
25. Dundálek, J.; Šnajdr, I.; Libánský, O.; Vrána, J.; Pocedič, J.; Mazúr, P.; Kosek, J. Zinc electrodeposition from flowing alkaline zincate solutions: Role of hydrogen evolution reaction. *J. Power Sources* **2017**, *372*, 221–226. [CrossRef]
26. Wang, K. Solutions for Dendrite Growth of Electrodeposited Zinc. *ACS Omega* **2020**, *5*, 10225–10227. [CrossRef]
27. Lu, W.; Xie, C.; Zhang, H.; Li, X. Inhibition of Zinc Dendrite Growth in Zinc-Based Batteries. *ChemSusChem* **2018**, *11*, 3996–4006. [CrossRef]
28. Hu, N.; Tao, J.; Tan, Y.; Song, H.; Huang, D.; Liu, P.; Chen, Z.; Yin, X.; Zhu, J.; Xu, J.; et al. Comprehensive Understanding of Steric-Hindrance Effect on the Trade-Off Between Zinc Ions Transfer and Reduction Kinetics to Enable Highly Reversible and Stable Zn Anodes. *Adv. Energy Mater.* **2024**, *14*, 2404018. [CrossRef]
29. Guo, W.; Tian, F.; Fu, D.; Cui, H.; Song, H.; Wang, C. High-Performance Aqueous Calcium Ion Batteries Enabled by Zn Metal Anodes with Stable Ion-Conducting Interphases. *Nano Lett.* **2024**, *24*, 12095–12101. [CrossRef]
30. Yang, Z.; Lv, C.; Li, W.; Wu, T.; Zhang, Q.; Tang, Y.; Shao, M.; Wang, H. Revealing the Two-Dimensional Surface Diffusion Mechanism for Zinc Dendrite Formation on Zinc Anode. *Small* **2021**, *18*, 2104148. [CrossRef]
31. Liu, Y.; Li, L.; Ji, X.; Cheng, S. Scientific Challenges and Improvement Strategies of Zn-Based Anodes for Aqueous Zn-Ion Batteries. *Chem. Rec.* **2022**, *22*, e202200114. [CrossRef]
32. Wang, X.; Sun, C.; Wu, Z.-S. Recent progress of dendrite-free stable zinc anodes for advanced zinc-based rechargeable batteries: Fundamentals, challenges, and perspectives. *SusMat* **2023**, *3*, 180–206. [CrossRef]
33. Du, W.; Zhang, Z.; Iacoviello, F.; Zhou, S.; Owen, R.E.; Jervis, R.; Brett, D.J.L.; Shearing, P.R. Observation of Zn Dendrite Growth via Operando Digital Microscopy and Time-Lapse Tomography. *ACS Appl. Mater. Interfaces* **2023**, *15*, 14196–14205. [CrossRef] [PubMed]
34. Li, Z.; Robertson, A.W. Electrolyte engineering strategies for regulation of the Zn metal anode in aqueous Zn-ion batteries. *Battery Energy* **2023**, *2*, 20220029. [CrossRef]
35. Liang, Y.; Qiu, M.; Sun, P.; Mai, W. Comprehensive Review of Electrolyte Modification Strategies for Stabilizing Zn Metal Anodes. *Adv. Funct. Mater.* **2023**, *33*, 2304878. [CrossRef]
36. Dai, H.; Sun, T.; Zhou, J.; Wang, J.; Chen, Z.; Zhang, G.; Sun, S. Unraveling chemical origins of dendrite formation in zinc-ion batteries via in situ/operando X-ray spectroscopy and imaging. *Nat. Commun.* **2024**, *15*, 8577. [CrossRef]
37. Hao, G.; Tian, X.; Chen, X.; Dong, Y.; Hao, C.; Li, C. Challenges and strategies for constructing stable zinc metal anodes in mild aqueous zinc-ion batteries. *Mater. Today Chem.* **2024**, *38*, 102059. [CrossRef]
38. He, C.; Gou, Q.; Hou, Y.; Wang, J.; You, X.; Yang, N.; Tian, L.; Xie, G.; Chen, Y. Growth and inhibition of zinc anode dendrites in Zn-air batteries: Model and experiment. *Chin. J. Chem. Eng.* **2024**, *67*, 268–281. [CrossRef]
39. Zhai, C.; Zhao, D.; He, Y.; Huang, H.; Chen, B.; Wang, X.; Guo, Z. Electrolyte Additive Strategies for Suppression of Zinc Dendrites in Aqueous Zinc-Ion Batteries. *Batteries* **2022**, *8*, 153. [CrossRef]
40. Nikolić, N.D.; Lović, J.D.; Maksimović, V.M.; Živković, P.M. Morphology and structure of electrolytically synthesized tin dendritic nanostructures. *Metals* **2022**, *12*, 1201. [CrossRef]

41. Popov, K.I.; Pavlović, M.G.; Maksimović, M.D.; Krstajić, S. The comparison of galvanostatic and potentiostatic copper powder deposition on platinum and aluminium electrodes. *J. Appl. Electrochem.* **1978**, *8*, 503–514. [CrossRef]
42. Xing, Z.; Huang, C.; Hu, Z. Advances and strategies in electrolyte regulation for aqueous zinc-based batteries. *Coord. Chem. Rev.* **2022**, *452*, 214299. [CrossRef]
43. Berube, L.P.; Esperance, G.L. A Quantitative Method of Determining of the Degree of Texture of Zinc Electrodeposits. *J. Electrochem. Soc.* **1989**, *136*, 2314–2315. [CrossRef]
44. Diggle, J.W.; Despic, A.R.; Bockris, J.O. Mechanism of dendritic electrocrystallization of zinc. *J. Electrochem. Soc.* **1969**, *116*, 1503–1514. [CrossRef]
45. Matsushita, M.; Sano, M.; Hayakawa, Y.; Honjo, H.; Sawada, Y. Fractal structures of zinc metal leaves grown by electrodeposition. *Phys. Rev. Lett.* **1984**, *53*, 286–289. [CrossRef]
46. Wheeler, A.A.; Murray, B.T.; Schaefer, R.J. Computation of dendrites using a phase field model. *Phys. D Nonlinear Phenom.* **1993**, *66*, 243–262. [CrossRef]
47. Cogswell, D.A. Quantitative phase-field modeling of dendritic electrodeposition. *Phys. Rev.* **2015**, *92*, 011301. [CrossRef]
48. Popov, K.I.; Nikolić, N.D. General Theory of Disperse Metal Electrodeposits Formation. In *Electrochemical Production of Metal Powders, Series: Modern Aspects of Electrochemistry*; Djokić, S.S., Ed.; Springer: New York, NY, USA, 2012; Volume 54, pp. 1–62.
49. Winand, R. Electrodeposition of Metals and Alloys–New Results and Perspectives. *Electrochim. Acta* **1994**, *39*, 1091–1105. [CrossRef]
50. Nikolić, N.D.; Maksimović, V.M.; Branković, G. Morphological and Crystallographic Characteristics of Electrodeposited Lead from the Concentrated Electrolyte. *RSC Adv.* **2013**, *3*, 7466–7471. [CrossRef]
51. Fu, H.; Xiong, L.; Han, W.; Wang, M.; Kim, Y.J.; Li, X.; Yang, W.; Liu, G. Highly active crystal planes-oriented texture for reversible high-performance Zn metal batteries. *Energy Storage Mater.* **2022**, *51*, 550–558. [CrossRef]
52. Wang, X.; Meng, J.P.; Lin, X.G.; Yang, Y.D.; Zhou, S.; Wang, Y.P.; Pan, A.Q. Stable zinc metal anodes with textured crystal faces and functional zinc compound coatings. *Adv. Funct. Mater.* **2021**, *31*, 2106114. [CrossRef]
53. Jia, X.X.; Liu, C.F.; Neale, Z.G.; Yang, J.H.; Cao, G.Z. Active materials for aqueous zinc ion batteries: Synthesis, crystal structure, morphology, and electrochemistry. *Chem. Rev.* **2020**, *120*, 7795–7866. [CrossRef] [PubMed]
54. Wu, S.; Hu, Z.; He, P.; Ren, L.; Huang, J.; Luo, J. Crystallographic engineering of Zn anodes for aqueous batteries. *eScience* **2023**, *3*, 100120. [CrossRef]
55. Zhou, M.; Guo, S.; Li, J.; Luo, X.; Liu, Z.; Zhang, T.; Cao, X.; Long, M.; Lu, B.; Pan, A.; et al. Surface-preferred crystal plane for a stable and reversible zinc anode. *Adv. Mater.* **2021**, *33*, 2100187. [CrossRef] [PubMed]

metals
MDPI

Article

Study on Microstructure and High-Temperature Mechanical Properties of Al-Mg-Sc-Zr Alloy Processed by LPBF

Zhihao Ren [1,2], Hao Zhang [2], Xuedao Shu [1,*], Haijie Xu [1], Siyuan Chen [1], Yaoyao Ding [3], Liwen Liang [2], Zixiang Qiu [2], Yang Yang [2] and Yongjian Zheng [2,*]

[1] Faculty of Mechanical Engineering and Mechanics, Ningbo University, Ningbo 315211, China; 2211090158@nbu.edu.cn (Z.R.); xuhaijie@nbu.edu.cn (H.X.); 2201090012@nbu.edu.cn (S.C.)

[2] Ningbo Institute of Materials Technology & Engineering, Chinese Academy of Sciences, Ningbo 315201, China; haozhang@nimte.ac.cn (H.Z.); liangliwen@nimte.ac.cn (L.L.); qizixiang@nimte.ac.cn (Z.Q.); beckyang@nimte.ac.cn (Y.Y.)

[3] China SciLong Lightweight Technology Co., Ltd., Ningbo 315336, China; dingyaoyao@zhongkexianglong.com

* Correspondence: shuxuedao@nbu.edu.cn (X.S.); yongjian.zheng@nimte.ac.cn (Y.Z.)

Abstract: Al-Mg-Sc-Zr alloy processed via laser powder bed fusion (LPBF) is poised for significant application in aerospace, where its high-temperature capabilities are paramount for the safety and longevity of engineered structures. This study offers a systematic examination of the alloy's high-temperature tensile properties in relation to its microstructure and precipitate phases, utilizing experimental approaches. The LPBF-processed Al-Mg-Sc-Zr alloy features a bimodal microstructure, with columnar grains in the melt pool's interior and equiaxed grains along its boundary, conferring exceptional properties. The application of well-calibrated processing parameters has yielded an alloy with an impressive relative density of 99.8%, nearly fully dense. Following a thermal treatment of 350 °C for 4 h, the specimens were subjected to tensile tests at both room and elevated temperatures. The data reveal that the specimens exhibit a tensile strength of 560.6 MPa and an elongation of 11.1% at room temperature. A predictable decline in tensile strength with rising temperature is observed: at 100 °C, 150 °C, 200 °C, and 250 °C; the respective strengths and elongations are 435.1 MPa and 25.8%, 269.4 MPa and 20.1%, 102.8 MPa and 47.9%, 54.0 MPa and 72.2%. These findings underpin the technical rationale for employing LPBF-processed Al-Mg-Sc-Zr alloy in aerospace applications.

Keywords: additive manufacturing; aluminum alloy; microstructure; high-temperature performance

1. Introduction

Laser powder bed fusion (LPBF) is an innovative additive manufacturing technique that integrates 3D modeling, material design, and automated control from various disciplines. This technology enables the rapid formation of parts from metal powders, eliminating the need for molds, machining, and assembly processes traditionally required in manufacturing. LPBF is a highly material-efficient process. By melting powder layer by layer, it minimizes material waste, which is particularly crucial for expensive and rare alloy materials. Moreover, LPBF technology significantly shortens production cycles and reduces tooling costs, making prototype design and small-batch production more economically feasible. In the aerospace sector, this facilitates rapid iteration and design optimization, accelerating the implementation of new technologies and materials. The characteristic of forming parts in a single step allows for weight reduction through integrated structural design, as demonstrated in recent studies [1,2]. LPBF-processed aluminum alloys, known for their high specific strength, excellent machinability, and superior corrosion resistance, are gaining significant traction in industries such as aerospace and automotive manufacturing due to their broad application prospects [3–6].

In recent years, an advanced Al-Mg-Sc-Zr alloy has emerged as an ideal candidate for LPBF. This alloy stands out as a multifunctional material, demonstrating exceptional repeatable printability and superior mechanical properties. Derived from an Al-Mg-based alloy modified with the addition of Sc and Zr elements, it has successfully been commercialized in aerospace and automotive industries, exemplified by Scalmalloy®—a material with the composition Al-4.5Mg-0.75Sc-0.35Zr-0.45Mn [7–10]. Current research indicates that LPBF-produced Al-Mg-Sc-Zr alloys can achieve a yield strength (YS) of up to 500 MPa at room temperature (RT), while maintaining an elongation at fracture (ε_f) of over 10%. The high strength of the Al-Mg-Sc-Zr alloy is attributed to its microstructure, which features a bimodal grain structure, known as a dual-phase structure. This consists of fine grains (FG) distributed along the melt pool boundaries and coarse grains (CG) within the melt pool interior. The fine grain regions are formed by the precipitation of Al_3(Sc, Zr) phases during solidification at the melt pool boundaries. These precipitates have lattice parameters similar to those of α-Al and a low lattice mismatch, making them ideal nucleation sites for Al grains, thus promoting heterogeneous nucleation [11]. This approach significantly mitigates the issue of cracking, providing effective nucleation sites for the crystallization of α-Al grains, thereby reducing the formation of columnar crystals [12,13]. The heterogeneous nucleation of columnar grains within the melt pool is attributed to the rapid cooling from temperatures above 800 degrees Celsius, which generates a substantial temperature gradient [14–16]. Existing research indicates that optimizing LPBF process parameters—such as laser power, scan speed, layer thickness, hatch spacing, powder layer thickness, and scan path—can significantly enhance the alloy's density and microstructure, thereby improving its mechanical properties [17–23]. The research demonstrates that Al-Mg-Sc-Zr alloys processed by the LPBF process possess exceptional mechanical performance and robust corrosion resistance under room-temperature conditions. Despite the extensive studies on LPBF-processed Al-Mg-Sc-Zr alloys, further research is needed to understand their tensile properties in high-temperature environments, particularly the microstructural evolution and changes in mechanical properties under varying temperature conditions.

Researching the high-temperature tensile properties of Al-Mg-Sc-Zr alloy processed by LPBF holds significant practical importance and potential industry impact. High-temperature environments are common in the aerospace sector, making it crucial to study the tensile properties of this alloy at elevated temperatures for designing and manufacturing highly reliable aerospace components. Furthermore, understanding the microstructural evolution and mechanical property changes of the alloy under different temperature conditions aids in optimizing LPBF process parameters, further enhancing material performance and manufacturing quality. This study systematically investigates the tensile property anisotropy of LPBF Al-Mg-Sc-Zr alloy in the as-built (AB) state and after annealing at 350 °C for 4 h (as-aged, AA). The 4 h annealing process is also considered as a heat treatment to alleviate residual stresses, which are a known issue for LPBF components. In light of this, the research characterizes the microstructure of the LPBF-formed Al-Mg-Sc-Zr alloy, performs tensile experiments at room and elevated temperatures. It examines the tensile properties and fracture morphology, establishing a relationship between microstructure, anisotropy, and high-temperature mechanical performance, which is significant for the application and advancement of LPBF-processed Al-Mg-Sc-Zr alloys.

2. Materials and Methods

2.1. Powder and LPBF Parameters

The Al-Mg-Sc-Zr powder utilized in this study was graciously supplied by Baohang Advanced Material Co., Ltd. (Nanchang, China) This powder was meticulously produced through gas atomization technology, a method specifically optimized for use with LPBF machinery in 3D printing applications. The scanning strategy employed adheres to the machine's default checkerboard pattern. A detailed composition of the material is presented in Table 1, which reveals that the measured content of each constituent within the Al-Mg-Sc-Zr powder falls well within the prescribed standard range. For the processing of the

Al-Mg-Sc-Zr powder, the HBD-450 3D printing equipment, provided by Hanbang Laser Technology Co., Ltd. (Zhongshan, China), was chosen. The specific printing parameters are delineated in Table 2. Additionally, the volume energy density (*VED*) is determined through the application of the formula presented below [24]:

$$VED = \frac{P}{V \times h \times t}$$

P is the laser power, *V* is the laser scanning speed, *h* is the hatching distance, and *t* is the layer thickness.

Table 1. Chemical composition of Al-Mg-Sc-Zr powder.

Elements	Mg	Sc	Zr	Mn	Si	Al
wt%	7.80	0.59	0.31	0.47	0.66	Bal

Table 2. Printing parameters of Al-Mg-Sc-Zr powder.

Laser Power	Scanning Speed	Layer Thickness	Hatching Distance	*VED*
400 W	1250 mm/s	50 μm	120 μm	53.33 J/mm^3

The particle size distribution of the powder ranges from 12 to 72 μm, with the specific diameters at 10%, 50%, and 90% cumulative distribution (d10, d50, d90) being 19.53 μm, 35.18 μm, and 57.68 μm, respectively. The morphology of the powder is depicted in Figure 1, revealing a high sphericity and smooth surface overall. However, minor instances of particle agglomeration and a small number of irregularly shaped particles are observed. Such irregularities can impair the powder's flowability, leading to uneven spreading and consequently reducing the density of the formed parts. Samples were annealed in a muffle furnace at 350 °C for 4 h, as illustrated in Figure 2. This heat treatment results in the supersaturation of Mn within the matrix and the formation of nanoscale Al_3Sc precipitates, which contribute to solution strengthening, grain refinement, and precipitation strengthening.

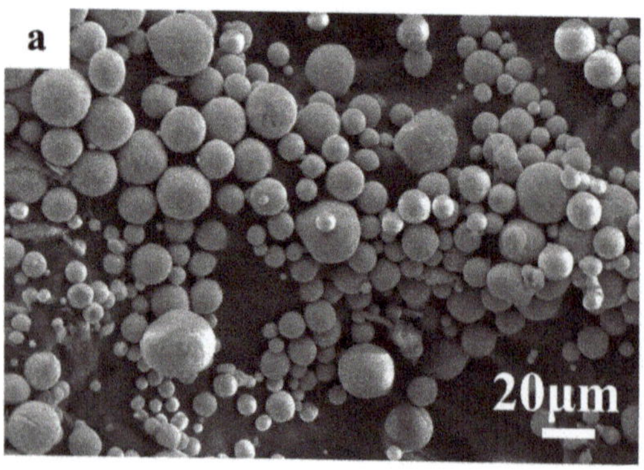

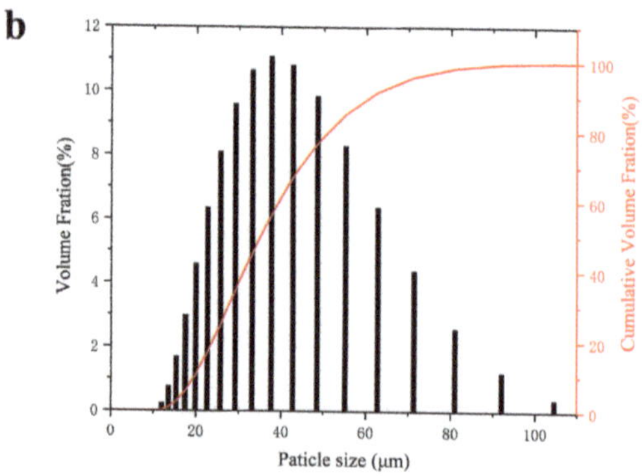

Figure 1. (**a**) The morphology of Al-Mg-Sc-Zr powder was observed by SEM; (**b**) the particle size of the powder was tested by a laser particle size analyzer.

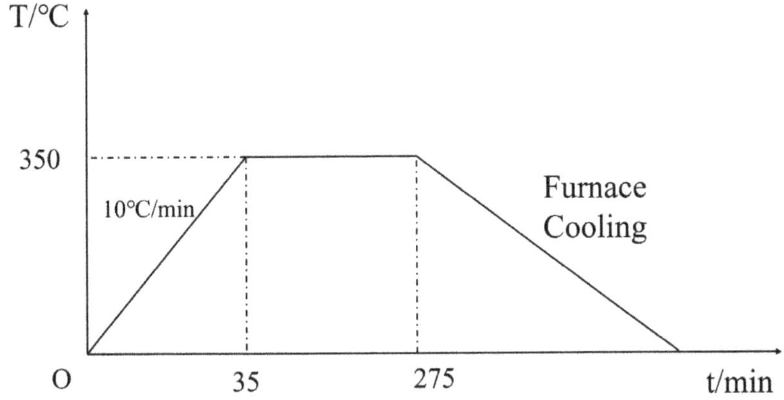

Figure 2. Heat treatment method of Al-Mg-Sc-Zr alloy.

2.2. Microstructure Characterization

The density of each sample was measured using the Archimedes drainage method. The formed samples were wire-cut and subsequently polished using progressively finer sandpapers to obtain standard metallographic specimens. These specimens were etched for 15 s with Keller's reagent (1 mL HF, 1.5 mL HCl, 2.5 mL HNO_3, and 95 mL H_2O) for surface morphology and microstructure observation using optical microscopy (OM, Axio Observer, Carl Zeiss AG, Oberkochen, Germany) and scanning electron microscopy (SEM, GEMINI 300, Carl Zeiss AG, Oberkochen, Germany). For electron backscatter diffraction (EBSD) analysis, the samples were electrolytically polished at −50 °C using an electrolyte solution composed of 5% perchloric acid and 95% ethanol. The grain size and crystallographic orientation data were extracted using AZtecCrystal software (v2.0, Oxford Instruments NanoAnalysis, High Wycombe, UK). In this study, high resolution transmission electron microscopy (HRTEM, TF20, FEI Company, Hillsboro, OR, USA) was used to examine the second-phase particles in the alloy's microstructure. Samples were prepared by wire-cutting 10 mm × 10 mm × 0.3 mm sections, which were then ground to a thickness of 150 µm using sandpaper. Subsequently, a 3 mm diameter disc was punched from the thinned section. The final stage of sample preparation involved electrolytic twin-jet polishing, using an electrolyte consisting of 30% nitric acid (HNO_3) and 70% methanol (CH_3OH), cooled to −30 °C. This process was carried out at an applied voltage of 20 V and a current of 120 mA. The samples were immediately rinsed with anhydrous ethanol after polishing. The prepared specimens were then subjected to HRTEM analysis.

2.3. Experimental Method of Mechanics

To test the tensile anisotropy of the Al-Mg-Sc-Zr alloy processed by LPBF, tensile specimens were categorized into H direction and Z direction based on the building direction, as shown in Figure 3a. The tensile tests were conducted according to the GB/T 228.2-2015 standard, with specimen dimensions illustrated in Figure 3b,c, applying an engineering strain rate of 7×10^{-5} s^{-1}. During high-temperature tensile tests, the specimens were preconditioned at the specified temperature for 30 min before loading. To ensure data reproducibility, each tensile test was performed three times, and the average value was taken.

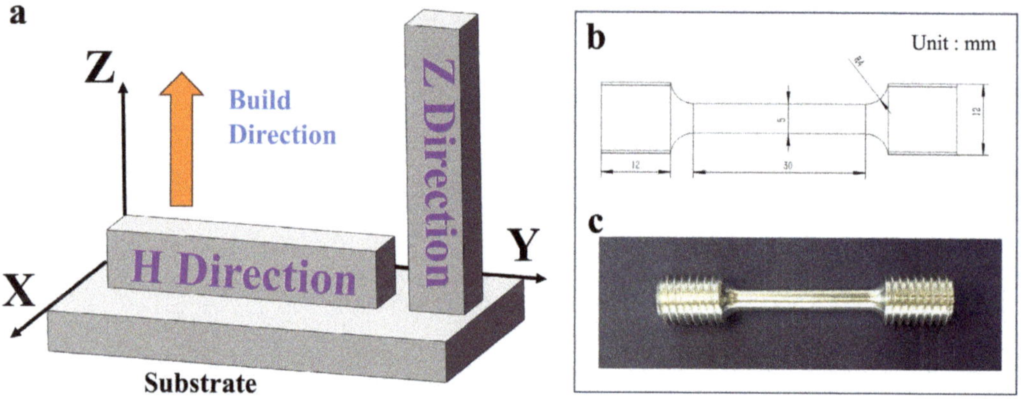

Figure 3. (**a**) Al-Mg-Sc-Zr tensile bar processing direction diagram; (**b**) drawings of Al-Mg-Sc-Zr tensile test rods; (**c**) Al-Mg-Sc-Zr tensile test bar.

3. Results and Discussion

3.1. Microstructure Observations

Figure 4 displays the optical micrographs (OM) of samples from the Al-Mg-Sc-Zr alloy. The relative density of the samples, determined by the Archimedes method, reached 99.8%. The microstructure showed no visible cracks, confirming the successful fabrication of high-density samples using LPBF technology. Figure 5 provides OM images of the etched Al-Mg-Sc-Zr alloy samples. In the XOY and XOZ planes, a distinct dark stripe was observed in the brighter regions, indicating the boundary of the melt pool, while the brighter regions represent the interior of the melt pool. The shape of the melt pool was unevenly distributed in the XOY and XOZ planes. In the XOY plane, the melt pool appeared as irregular ellipses. In the YOZ plane, the melt pool exhibited a fan shape, characterized by a regular, fish-scale-like arrangement. Measurements of the observed surface indicated that the size of the melt pools ranged from 20 μm to 120 μm.

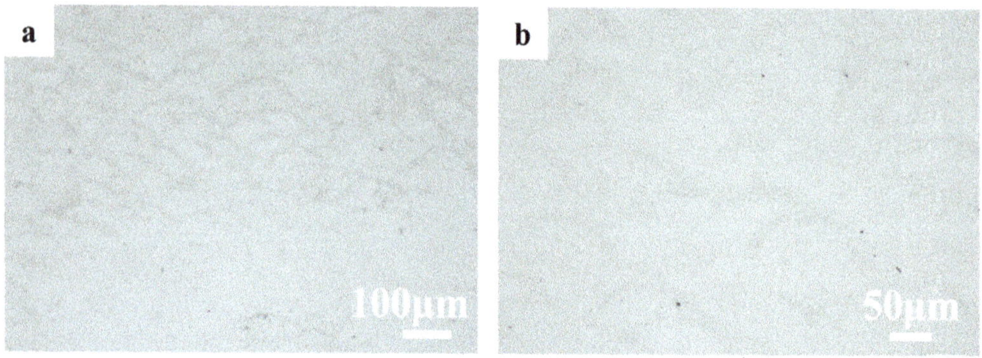

Figure 4. Morphology of Al-Mg-Sc-Zr alloy by OM. (**a**) 100× magnification; (**b**) 200× magnification.

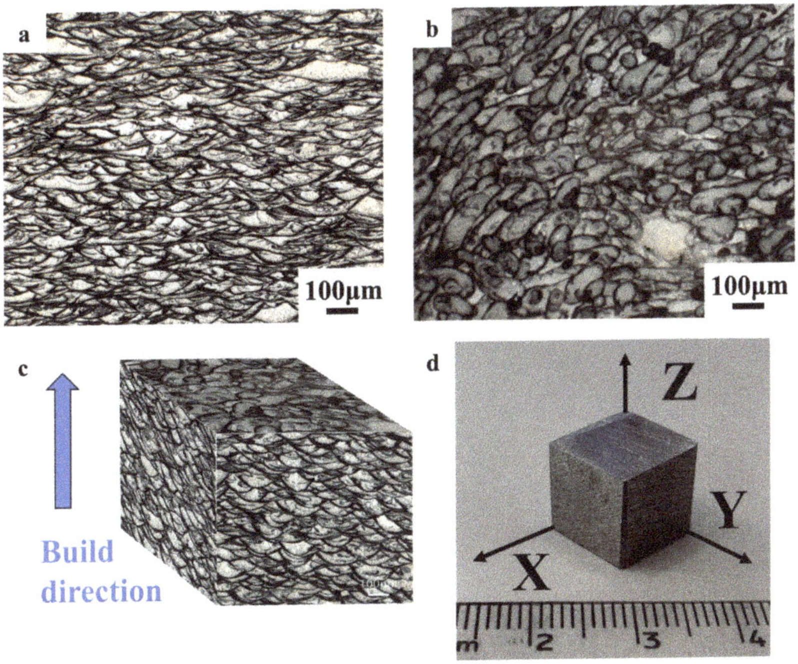

Figure 5. (**a**) Morphology of YZ plane after corrosion by OM; (**b**) morphology of XY plane after corrosion by OM; (**c**) 3d metallographic diagram of test block; (**d**) Al-Mg-Sc-Zr test block.

Figure 6a displays the X-ray diffraction (XRD, D8 Advance, Bruker Corporation, Karlsruhe, Germany) patterns of the pristine powder and the sample produced by the laser powder bed fusion (LPBF) technique. The XRD analysis confirms that the α-Al phase constitutes the principal crystalline phase in both the powder and the LPBF-processed samples. Notably, no Mg-containing phases or other phases resulting from processing were identified in the XRD patterns, possibly due to their low abundance below the detection capabilities of XRD or the rapid scanning speed of the XRD equipment, which could hinder the detection of these phases. A significant observation is the increased intensity of the diffraction peak on the (111) crystal plane in the LPBF-processed sample when compared to the original powder. This suggests a pronounced preferred orientation on the (111) plane, likely a consequence of the thermal processes and crystallographic growth during LPBF, which can significantly affect the resulting microstructure and material properties. Figure 6b offers a detailed perspective of the XRD pattern between 37.5° and 39°. It is evident that the (111) diffraction peak of the α-Al phase in the sample prepared by LPBF has undergone a leftward shift relative to the standard α-Al peak at $2\theta = 38.446°$. This shift results in the (111) diffraction peaks aligning with lower 2θ values [25]. As dictated by Bragg's law, such a displacement to the left suggests an increase in the lattice parameters. The rapid solidification process characteristic of LPBF does not allow for timely diffusion of the majority of Mg atoms. These atoms are instead quickly assimilated into the α-Al matrix as solutes, causing substantial lattice distortion. This distortion is responsible for the observed leftward shift of the (111) peak.

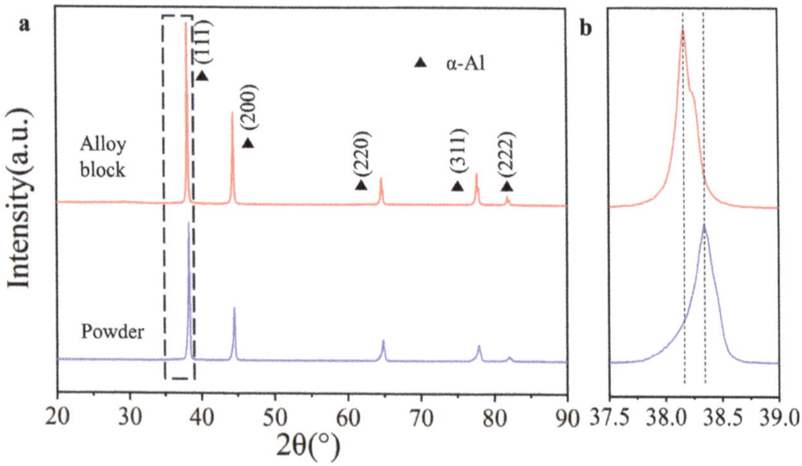

Figure 6. (**a**) X-ray diffraction patterns of the original powder and LPBF alloy block; (**b**) 2θ amplification region from 37.5° to 39.0°.

To further investigate the microstructural properties of the samples processed by LPBF, such as grain size, shape, and texture, EBSD analysis was conducted. Figure 7 shows the EBSD inverse pole figures and pole figures of the XOY and XOZ planes of the alloy. As shown in Figure 7a, the ultrafine equiaxed grains of the Al-Mg-Sc-Zr alloy are distributed around the periphery of the melt pool on the XOZ plane, while elongated grains are mainly found within the melt pool. Figure 7b reveals that the columnar crystals in the melt pool exhibit a preferred orientation along <100>. Additionally, the melt pool depth is approximately 150 μm, significantly exceeding the alloy's 50 μm layer thickness. This is attributed to the excessively high melt pool temperature during the LPBF process, which causes remelting of previously solidified material during subsequent layer scans [26,27].

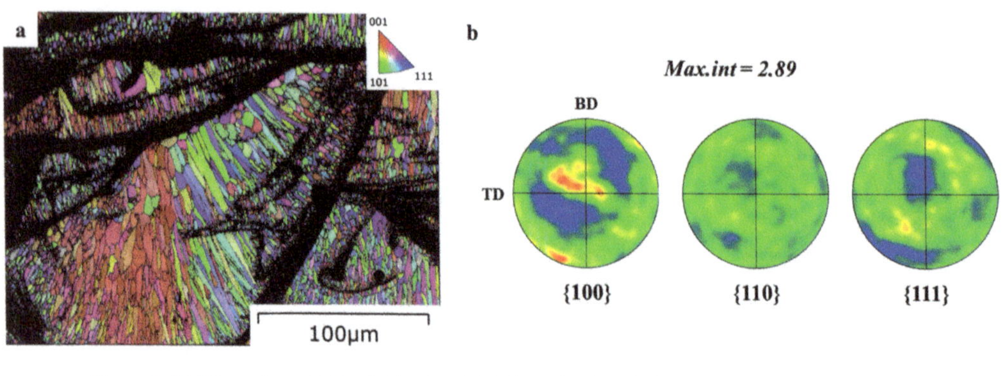

Figure 7. EBSD results of Al-Mg-Sc-Zr alloy. (**a**) IPF graph; (**b**) pole figure.

The SEM images in Figure 8 offer insights into the microstructure of the Al-Mg-Sc-Zr alloy crafted by LPBF. Figure 8a displays large columnar crystals in the heart of the molten pool, with several black corrosion pits evident at the pool's edges. A high-magnification SEM image, Figure 8b, captures nanoscale white precipitated phase particles within and encircling the corrosion pits at the pool's edge. These particles are postulated to be nano-Al_3(Sc, Zr), as per the analysis in Reference [28]. The temperature gradients within the molten pool, under suitable process parameters, promote the precipitation of these thermally stable Al_3(Sc, Zr) particles [29]. As such, these particles precipitate stably and accumulate at the periphery of the molten pool during the repeated cycles of melting and solidification.

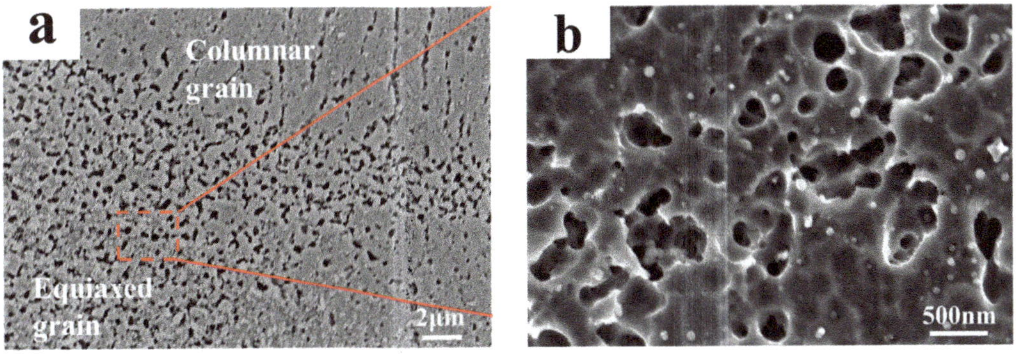

Figure 8. SEM images of Al-Mg-Sc-Zr alloy processed by LPBF. (**a**) Entire melt pool; (**b**) melt pool boundary.

The HRTEM-BF-S images in Figure 9, complemented by EDS spectra, provide a detailed examination of the fine grain zone in the 3D printed samples, highlighting the concentration of Sc and Zr elements around this area. The precipitated secondary phase particles are identified as Al_3(Sc, Zr) based on HRTEM and SAED analyses. The LPBF process involves a laser scanning a powder bed, with each layer being elevated and melted in sequence to achieve a layered scanning and melting effect. The formation of a molten pool upon laser interaction with Al powder results in the distinctive fish-scale pattern observable in the alloy's microstructure (as shown in Figure 5). The cooling rate of these molten pools is exceptionally high, between 10^3 and 10^6 K/s [30]. As the molten pool cools, solidification initiates at the edges due to uneven heat dissipation. The hypereutectic elements Sc and Zr form a fine, dispersed $L1_2$ phase with Al, known as the Al_3(Sc, Zr) precipitate. This phase is coherent with the Al matrix [31] and is prevalent at the fine grain zone's grain boundaries, where it serves to pin the boundaries, restrict dislocation movement, and also acts as an effective nucleation site for grain refinement [32]. The solidification process at the molten pool boundary leads to the consumption of the Sc element, thereby eliminating potential nucleation points within the molten pool. Additionally, the nth layer's molten pool is subject to the influence of neighboring molten pools (as shown in Figure 7, where there is a preference for columnar crystal orientation) and the subsequent n + 1th layer, which collectively restricts solidification within the molten pool. With temperature gradients diminishing from a high to a lower range, solidification within the molten pool initiates at its boundary. The depletion of Sc at the boundary, coupled with a substantial temperature gradient, results in solidification occurring counter to the direction of heat transfer, thus giving rise to columnar crystals. This explains the bimodal microstructure characteristic of the LPBF-processed Al-Mg-Sc-Zr alloy. As shown in Figure 10, the dispersion of nano-sized Al_3(Sc, Zr) precipitates in the alloy's fine-grained region enhances its strength. The Hall–Petch strengthening model [33] dictates that the strength increment from grain boundaries is related to the inverse square root of the grain size, formulated as $\sigma GB \propto d\hat{}(-1/2)$. Based on reports in the existing body of literature [34], the alloy's strengthening mechanisms are considered to involve dispersion strengthening and the refinement of grain size.

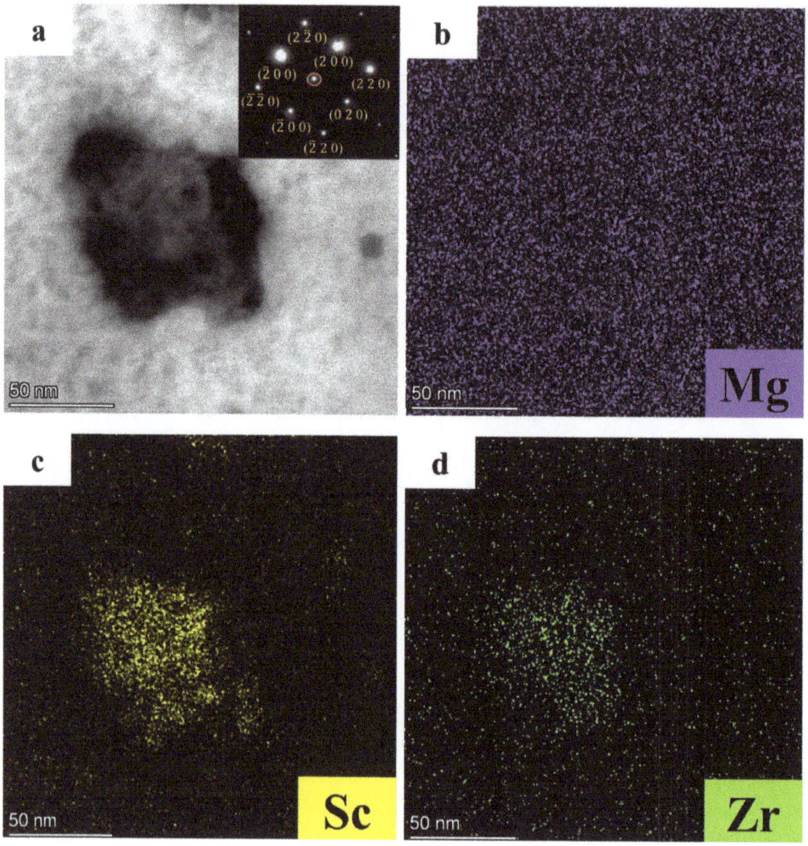

Figure 9. (**a**) HRTEM-BF-S images and selected-area electron diffraction pattern corresponding to the image; (**b–d**) corresponding EDS elemental distribution maps.

Figure 10. The fine grain zone at the boundary of the molten pool was observed under HRTEM.

3.2. Mechanical Properties

Tensile testing was performed on specimens AB in their as-built state and AA in their as-aged state, under both ambient and elevated temperature conditions. The resulting stress–strain curves are displayed in Figure 11, with 'H' indicating the specimens oriented perpendicular to the build direction and 'Z' indicating those oriented parallel. The mechanical properties, including yield strength (YS), ultimate tensile strength (UTS), and elongation, are summarized in Table 3. The Al-Mg-Sc-Zr alloy fabricated using the LPBF technique demonstrates a higher tensile strength compared to the commercial aluminum alloy sheets (5B70), with a measured value of 421 MPa [35]. This enhanced mechanical performance can be attributed to the crack-free nature and the pronounced grain refinement observed in the microstructure of the LPBF-processed material. In Table 3 and Figure 11, we observe that the elongation of AA-H at 150 °C is lower than that at 100 °C. This may be due to the fact that at 150 °C, internal microvoids and cracks in the material are more likely to propagate, leading to a significant decrease in elongation. At 200 °C, despite the higher temperature, the material likely undergoes more plastic deformation, and the propagation mechanism of microvoids and cracks differs from that at 150 °C.

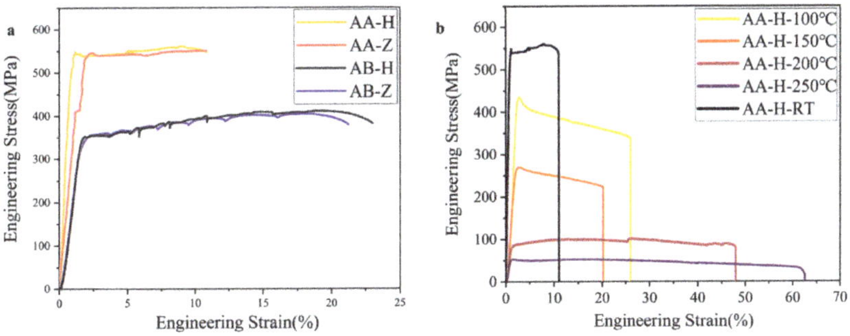

Figure 11. Al-Mg-Sc-Zr stress–strain curve. (**a**) Tensile test of AA specimens and AB specimens at room temperature; (**b**) tensile test of AA specimens at high temperature.

Table 3. Al-Mg-Sc-Zr tensile test data.

	YS (MPa)	UTS (MPa)	ELO (%)	E (GPa)
AB-Z-RT	321	404	21.3	69.9
AB-H-RT	352	412	23.6	72.2
AA-Z-RT	527	551	11.2	70.7
AA-H-RT	535	560	11.1	65.5
AA-H-100 °C	417	435	25.8	63.6
AA-H-150 °C	214	269	20.1	62.1
AA-H-200 °C	77	102	47.9	38.5
AA-H-250 °C	52	54	72.2	36.2

Fracture usually occurs in the weakest or most stressed part of the material, and the fracture morphology records the relevant information of the fracture process. By analyzing the deformation degree, position, color, precipitates, particle size, and pattern of the fracture, the nature of the fracture can be determined, the cause of the failure can be analyzed, and the mechanism of the fracture can be further studied [36]. In order to better understand the fracture mechanism, the fracture surface of Al-Mg-Sc-Zr samples prepared by LPBF after tensile was observed by SEM. From the macroscopic view, it can be found that the dimples of different sizes can be found in the fracture of the samples, so it can be determined that the tensile fracture of the alloy is a typical ductile fracture. With the increase in tensile temperature, the fracture of the sample is more uneven, which is related to the increase in elongation. As depicted in Figure 12a, the fracture surface morphology of the AA specimen exhibits characteristics akin to columnar grain structures at the crack origin site (a). Such features are exclusively observed on the fracture crack surfaces of specimens following heat treatment. This can be attributed to two primary factors: The material's bimodal microstructure, characterized by a varied grain size distribution, results in an inhomogeneous stress distribution. The equiaxed grain areas, reinforced by the pinning effect of Al_3Sc particles at the grain boundaries, demonstrate greater strength than the columnar grain regions. Consequently, crack initiation is more prone to occur in the columnar grain areas. The existence of larger grains within the bimodal microstructure contributes to a reduced crack propagation velocity, which in turn, augments the material's fracture toughness. Figure 12c,d, representing the fracture surfaces at temperatures of 100 degrees and 200 degrees, respectively, corroborate the presence of this phenomenon.

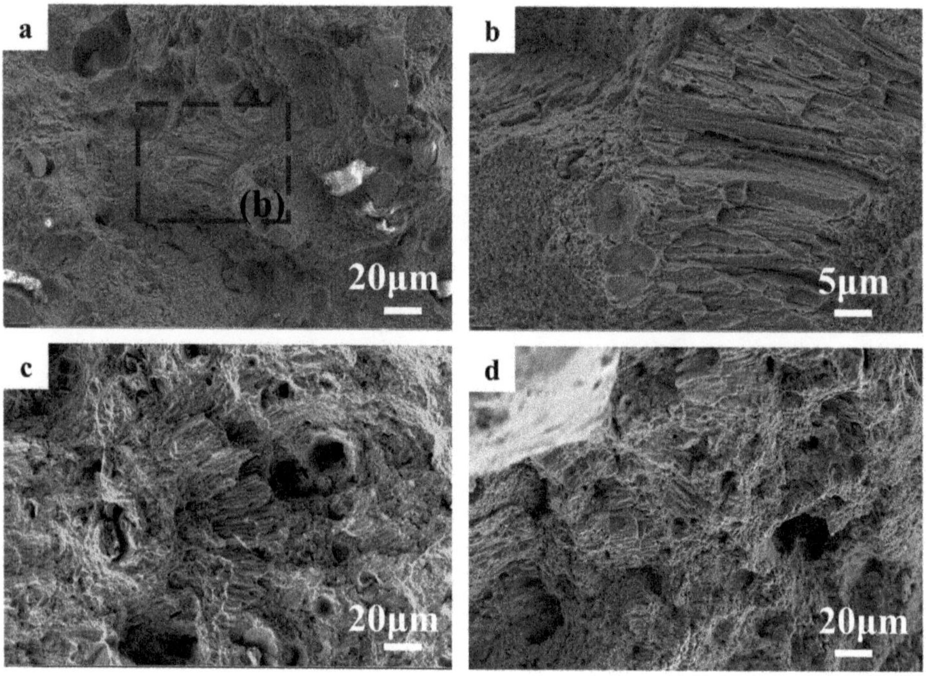

Figure 12. Tensile fracture of test rod. (**a**) AA-RT; (**b**) regional enlargement of (**a**); (**c**) 100 °C; (**d**) 200 °C.

Ductile fractures are generally characterized by three distinct regions: the fiber zone, which is also known as the goose down zone, the radiation zone, and the shear lip zone. These zones are the key components of fracture morphology [37]. As depicted in Figure 13a,c, the fiber zone of the tensile fracture is examined across a range of temperatures, showing an increase in irregularity with rising temperature. The crack propagation within this zone is slower in comparison to other areas. The fracture surface features a jagged pattern punctuated by holes, which at a microscopic level, is composed of many fibrous 'mini peaks', each with a 45° inclination to the tensile axis. This inclination indicates that the fiber zone is a result of the progressive enlargement and merging of microcracks during plastic deformation [38]. The dimples on the fracture surface are a consequence of dislocation slip, with the second-phase particles acting as the main impetus. Figure 13b,d show the dimple morphology on the fracture surfaces at 100 °C and 200 °C, respectively. The dense distribution of dimples indicates that the material exhibits high ductility. The dimples at 200 °C are larger than those at 100 °C, suggesting that the specimens underwent greater plastic deformation before fracture. This observation is consistent with the measured elongation results. The size and shape of these dimples are correlated with the material's properties, including its strength, the dimensions, form, and dispersion of the second-phase particles, as well as the stress conditions, encompassing both the nature and magnitude of the stress applied [39]. Figure 9 displays the secondary phase particles located at the grain boundaries in the equiaxed zone, as detected by EDS. The elemental analysis confirms their enrichment with Scandium (Sc) and Zirconium (Zr). Drawing from Wang's characterization [40], these particles are identified as the primary $Al_3(Sc, Zr)$ phase, which is directly responsible for the dimple formation observed in the tensile fracture surfaces of Al-Mg-Sc-Zr alloy systems. In alignment with dislocation theory [41], it is around these $Al_3(Sc, Zr)$ particles that dislocation loops are formed, contributing to the material's ductile fracture behavior. As depicted in Figure 14, in the absence of external forces, dislocation loops are in equilibrium, held in check by the repulsion from $Al_3(Sc, Zr)$ particles and the stress of dislocation stacking. When an external force is sufficiently large, dislocations around the $Al_3(Sc, Zr)$ particles are set into motion once more, moving toward the particles. The formation of microvoids occurs when the elastic strain energy that has built up in front of the dislocation loop reaches a level that can overcome the interfacial bonding forces between the $Al_3(Sc, Zr)$ particles and the aluminum matrix. The formation of microvoids significantly alleviates the repulsive forces on dislocations, leading to a surge of dislocations toward the microvoids under external force, which in turn, drives their enlargement. At this juncture, the dislocation sources that were quiescent behind the dislocation loop become re-energized as the constraint from dislocation accumulation is lifted, initiating new dislocation loops that migrate incessantly toward the microvoids. This results in a swift and unstable enlargement and coalescence of the microvoids. Dislocations, capable of amassing on diverse slip planes, may induce the formation of microvoids through their movement along one or several planes. Concurrently, dislocations on other slip planes are drawn to the microvoids, further promoting their expansion [42].

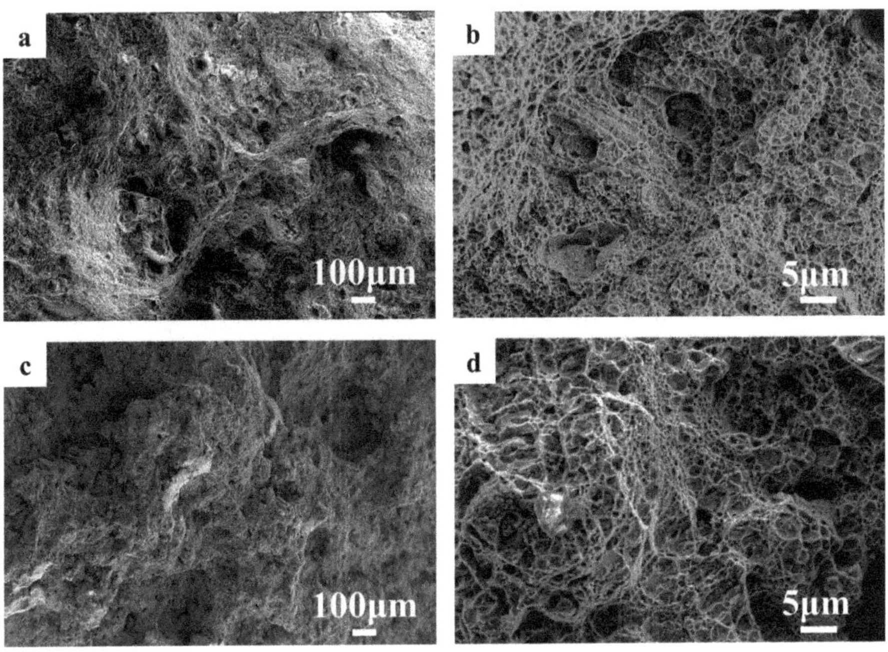

Figure 13. Tensile fracture of AA-H test rod. (**a**) The fiber zone of fracture at 100 °C; (**b**) dimple morphology of fracture at 100 °C; (**c**) the fiber zone of fracture at 200 °C; (**d**) dimple morphology of fracture at 200 °C.

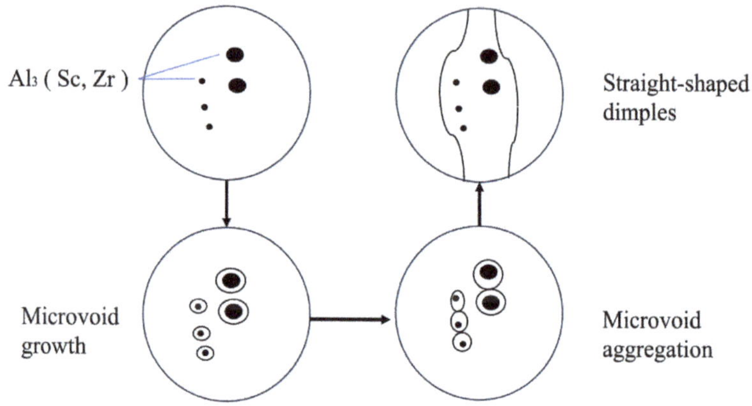

Figure 14. Schematic diagram of straight-shaped dimples processed by Al_3(Sc, Zr) particles.

4. Conclusions

The microstructure, room-temperature and high-temperature tensile properties and fracture morphology of Al-Mg-Sc-Zr samples prepared by laser powder bed fusion were studied. The main results are as follows:

- The microstructure is characterized by neatly aligned melt pools, with the boundaries marked by fine equiaxed grains and the interiors by larger equiaxed grains, creating a distinct bimodal distribution. EBSD analysis indicates an even dislocation distribution across the alloy, devoid of any textural preference.
- Room-temperature tensile tests demonstrated that the heat-treated specimens achieved ultimate tensile strengths up to 560.6 MPa, coupled with an elongation of 11.1%. Tensile strengths for both untreated and treated samples were modestly higher in the H direction than in the Z direction, with the discrepancy being less than 5%. A progressive decrease in strength with increasing test temperature was observed, with values at 100 °C being 435.1 MPa and 25.8%, at 150 °C being 269.4 MPa and 20.1%, at 200 °C being 102.8 MPa and 47.9%, and at 250 °C being 54.0 MPa and 72.2%, respectively. This trend signifies a trade-off between strength and ductility with rising temperature, with the material exhibiting greater plasticity at elevated temperatures.
- Fracture surface analyses from tensile tests across the temperature range disclosed an abundance of dimples, indicative of ductile fracture behavior. The dimples are mainly attributed to the presence of the second-phase Al_3(Sc, Zr) particles. Upon the application of substantial external forces, dislocations surrounding these particles become active again, leading to fracture when the forces surpass the interfacial bond strength between the particles and the aluminum matrix.

Author Contributions: Conceptualization: Z.R. and H.X. Methodology: H.Z. and Y.D. Software: S.C. and Y.Y. Validation: S.C. Formal Analysis: H.Z. and L.L. Investigation: X.S. and Y.Z. Resources: Y.Y. Data Curation: L.L. and Y.D. Writing—Original Draft Preparation: Z.R. and Y.Z. Writing—Review and Editing: Z.R. and Y.Z. Visualization: Z.Q. Supervision: H.X., and X.S. Project Administration: Z.Q. Funding Acquisition: H.X. and X.S. All authors have read and agreed to the published version of the manuscript.

Funding: This research was funded by "Pioneer" and "Leading Goose" R&D Program of Zhejiang, China (grant number: 2024C01121); Yongjiang talent introduction programmer (2021A-112-G;2022A-202-G); "3315 Project" innovation team (2020A-27-C); international partnership program of the Chinese Academy of Sciences, grant no. 181GJHZ2023132MI.

Data Availability Statement: The original contributions presented in the study are included in the article, further inquiries can be directed to the corresponding authors.

Conflicts of Interest: Author Yaoyao Ding was employed by the company China SciLong Lightweight Technology Co., Ltd. The remaining authors declare that the research was conducted in the absence of any commercial or financial relationships that could be construed as a potential conflict of interest.

References

1. Li, X.; Li, G.; Chen, H.; Zhang, M.; Xi, L. Novel approach to additively manufacture high-strength Al alloys by laser powder bed fusion through addition of hybrid grain refiners. *Addit. Manuf.* **2021**, *48*, 102400. [CrossRef]
2. Han, D.; Xiao, Y.; Chen, H.; Bian, Z.; Wu, Y.; Yang, H.; Wang, H.; Wang, H. Design, microstructure and thermal stability of a novel heat-resistant Al-Fe-Ni alloy manufactured by selective laser melting. *J. Alloys Compd.* **2021**, *885*, 160949.
3. Li, Q.; Li, G.; Lin, X.; Zhu, D.; Jiang, J.; Shi, S.; Liu, F.; Huang, W.; Vanmeensel, K. Development of a high strength Zr/Sc/Hf-modified Al-Mn-Mg alloy using laser powder Bed fusion: Design of a heterogeneous microstructure incorporating synergistic multiple strengthening mechanisms. *Addit. Manuf.* **2022**, *57*, 102967. [CrossRef]
4. Kotadia, H.R.; Gibbons, G.; Das, A.; Howes, P.D. A review of laser powder Bed fusion additive manufacturing of aluminum alloys: Microstructure and properties. *Addit. Manuf.* **2021**, *46*, 102155.
5. Lim, C.H.; Li, H.; Krishnan, M.; Chen, K.; Li, J. Novel method of residual stress reduction for AlSi10Mg manufactured using selective laser melting without compromise of mechanical strength. *Virtual Phys. Prototyp.* **2023**, *18*, 4. [CrossRef]
6. Tang, Z.; Wei, Q.; Gao, Z.; Yang, H.; Wang, A.; Wan, L.; Luo, C.; Wu, Y.; Wang, H.; Wang, H. 2000W blue laser directed energy deposition of AlSi7Mg: Process parameters, molten pool characteristics, and appearance defects. *Virtual Phys. Prototyp.* **2023**, *18*, 3. [CrossRef]
7. Palm, F.; Leuschner, R.; Schubert, T.; Kieback, B. Scalmalloy®= A unique high strength ALMgSc type material concept processed by innovative technologies for aerospace applications. In Proceedings of the World Powder Metallurgy Congress and Exhibition, World PM 2010, Florence, Italy, 10–14 October2010.
8. Kuo, C.N.; Peng, P.C.; Liu, D.H.; Chao, C.Y. Microstructure evolution and mechanical property response of 3D-Printed Scalmalloy with different heat-treatment times at 325 °C. *Metals* **2021**, *11*, 555. [CrossRef]
9. Schmidtke, K.; Palm, F.; Hawkins, A.; Emmelmann, C. Process and mechanical properties: Applicability of a scandium modified Al-alloy for laser additive manufacturing. *Phys. Procedia* **2011**, *12*, 369–374. [CrossRef]
10. Martucci, A.; Aversa, A.; Manfredi, D.; Bondioli, F.; Biamino, S.; Ugues, D.; Lombardi, M.; Fino, P. Low-power laser powder bed fusion processing of Scalmalloy®. *Materials* **2022**, *15*, 3123. [CrossRef]
11. Li, D.; Wu, Y.; Geng, Z.; Zhang, J.; Chen, C.; Liu, X.; Liu, Y.; Zhou, K. High strength Al−Mg−Sc−Zr alloy with heterogeneous grain structure and intragranular precipitation produced by laser powder bed fusion. *J. Alloys Compd.* **2023**, *939*, 168722. [CrossRef]
12. Jägle, E.A. Alloy Design for Additive Manufacturing. In *DARE Annual Workshop 2016*; University of Sheffield: Sheffield, UK, 2016.
13. Carluccio, D.; Bermingham, M.J.; Zhang, Y.; StJohn, D.H.; Yang, K.; Rometsch, P.A.; Wu, X.; Dargusch, M.S. Grain refinement of laser remelted Al-7Si and 6061 aluminium alloys with Tibor® and scandium additions. *J. Manuf. Process.* **2018**, *35*, 715–720. [CrossRef]
14. Spierings, A.B.; Dawson, K.; Heeling, T.; Uggowitzer, P.J.; Schäublin, R.; Palm, F.; Wegener, K. Microstructural features of Sc- and Zr-modified Al-Mg alloys processed by selective laser melting. *Mater. Des.* **2017**, *115*, 52–63. [CrossRef]
15. Yang, K.V.; Shi, Y.; Palm, F.; Wu, X.; Rometsch, P. Columnar to equiaxed transition in Al-Mg(-Sc)-Zr alloys produced by selective laser melting. *Scr. Mater.* **2018**, *145*, 113–117. [CrossRef]
16. Spierings, A.B.; Dawson, K.; Kern, K.; Palm, F.; Wegener, K. SLM-processed Sc- and Zr- modified Al-Mg alloy: Mechanical properties and microstructural effects of heat treatment. *Mater. Sci. Eng. A* **2017**, *701*, 264–273. [CrossRef]
17. Ekubaru, Y.; Gokcekaya, O.; Ishimoto, T.; Sato, K.; Manabe, K.; Wang, P.; Nakano, T. Excellent strength–ductility balance of Sc-Zr-modified Al–Mg alloy by tuning bimodal microstructure via hatch spacing in laser powder bed fusion. *Mater. Des.* **2022**, *221*, 110976. [CrossRef]
18. Cordova, L.; Bor, T.; de Smit, M.; Carmignato, S.; Campos, M.; Tinga, T. Effects of powder reuse on the microstructure and mechanical behaviour of Al–Mg–Sc–Zr alloy processed by laser powder bed fusion (LPBF). *Addit. Manuf.* **2020**, *36*, 101625. [CrossRef]
19. Wang, Z.; Lin, X.; Wang, J.; Kang, N.; Hu, Y.; Wang, D.; Li, H.; Huang, W.; Pan, F. Remarkable strength-impact toughness conflict in high-strength Al-Mg-Sc-Zr alloy fabricated via laser powder bed fusion additive manufacturing. *Addit. Manuf.* **2022**, *59*, 103093. [CrossRef]
20. Zhang, H.; Zhang, L.C.; Liu, H.; Niu, X.; Lam, M.C.; Zhang, W.; Jin, X.; Chu, F.; Wu, X.; Cao, S. Strong and ductile Al–Mn–Mg–Sc–Zr alloy achieved in fabrication-rate enhanced laser powder bed fusion. *Virtual Phys. Prototyp.* **2023**, *18*, e2250769. [CrossRef]
21. Zhou, Z.; Chen, C.; Guan, J.; Yuan, K. Study on the Effect of Remelt Scanning and Heat Treatment Processes on the Structure and Properties of Laser Powder Bed Fusion Formed Al-Mg-Sc-Zr Alloys. *J. Phys.* **2023**, *2566*, 012085. [CrossRef]
22. Zhang, Z.; Sun, J.; Wu, J.; Chen, Z.; Zhang, B.; Li, Z.; Li, X.; Zhang, B. Anisotropic response in corrosion behavior of laser powder bed fusion Al-Mn-Mg-Sc-Zr alloy. *Corros. Sci.* **2022**, *208*, 110634. [CrossRef]
23. Schliephake, D.; Lopes, C.; Eggeler, Y.M.; Chen, H.; Freudenberger, J.; Bayoumy, D.; Huang, A.J.; Kauffmann, A. Improved work hardening capability and ductility of an additively manufactured and deformed Al-Mn-Mg-Sc-Zr alloy. *J. Alloys Compd.* **2022**, *924*, 166499. [CrossRef]

24. Read, N.; Wang, W.; Essa, K.; Attallah, M.M. Selective laser melting of AlSi10Mg alloy: Process optimisation and mechanical properties development. *Mater. Des.* **2015**, *65*, 417–424. [CrossRef]
25. Zou, Y.; Tan, C.; Qiu, Z.; Ma, W.; Kuang, M.; Zeng, D. Additively manufactured SiC-reinforced stainless steel with excellent strength and wear resistance. *Addit. Manuf.* **2021**, *41*, 101971. [CrossRef]
26. Zou, T.; Mei, S.; Chen, M. Microstructure and Electrochemical Corrosion Properties of AlMgScZr Alloys Fabricated Using Selective Laser Melting. *Zhongguo Jiguang* **2023**, *50*, 85–94.
27. Spierings, A.B.; Dawson, K.; Uggowitzer, P.J.; Wegener, K. Influence of SLM scan-speed on microstructure, precipitation of Al_3Sc particles and mechanical properties in Sc-and Zr-modified Al-Mg alloys. *Mater. Des.* **2018**, *140*, 134–143. [CrossRef]
28. Zhang, H.; Gu, D.; Yang, J.; Dai, D.; Zhao, T.; Hong, C.; Gasser, A.; Poprawe, R. Selective laser melting of rare earth element Sc modified aluminum alloy: Thermodynamics of precipitation behavior and its influence on mechanical properties. *Addit. Manuf.* **2018**, *23*, 1–12. [CrossRef]
29. Deng, Y.; Peng, B.; Xu, G.; Pan, Q.; Yin, Z.; Ye, R.; Wang, Y.; Lu, L. Effects of Sc and Zr on mechanical property and microstructure of tungsten inert gas and friction stir welded aerospace high strength Al–Zn–Mg alloys. *Mater. Sci. Eng. A* **2015**, *639*, 500–513. [CrossRef]
30. Sing, S.; Wiria, F.; Yeong, W. Selective laser melting of titanium alloy with 50 wt% tantalum: Effect of laser process parameters on part quality. *Int. J. Refract. Met. Hard Mater.* **2018**, *77*, 120–127. [CrossRef]
31. Rometsch, P.; Zhong, H.; Nairn, K.M.; Jarvis, T.; Wu, X. Characterization of a laser-fabricated hypereutectic Al–Sc alloy bar. *Scr. Mater.* **2014**, *87*, 13–16. [CrossRef]
32. Li, G.; Zhao, N.; Liu, T.; Li, J.; He, C.; Shi, C.; Liu, E.; Sha, J. Effect of Sc/Zr ratio on the microstructure and mechanical properties of new type of Al–Zn–Mg–Sc–Zr alloys. *Mater. Sci. Eng. A* **2014**, *617*, 219–227. [CrossRef]
33. Hansen, N. Hall–Petch relation and boundary strengthening. *Scr. Mater.* **2004**, *51*, 801–806. [CrossRef]
34. Li, R.; Wang, M.; Li, Z.; Cao, P.; Yuan, T.; Zhu, H. Developing a high-strength Al-Mg-Si-Sc-Zr alloy for selective laser melting: Crack-inhibiting and multiple strengthening mechanisms. *Acta Mater.* **2020**, *193*, 83–98. [CrossRef]
35. Shen, J.; Chen, B.; Wan, J.; Shen, J.; Li, J. Effect of annealing on microstructure and mechanical properties of an Al–Mg-Sc-Zr alloy. *Mater. Sci. Eng. A* **2022**, *838*, 142821. [CrossRef]
36. Bayoumy, D.; Schliephake, D.; Dietrich, S.; Wu, X.H.; Zhu, Y.M.; Huang, A.J. Intensive processing optimization for achieving strong and ductile Al-Mn-Mg-Sc-Zr alloy produced by selective laser melting. *Mater. Des.* **2021**, *198*, 109317. [CrossRef]
37. Bathias, C. Application of fracture mechanics to aluminum alloys selection. *Eng. Fract. Mech.* **1978**, *10*, 267–282. [CrossRef]
38. Kucharíková, L.; Tillová, E.; Chalupová, M.; Uhríčik, M.; Pastierovičová, L.; Belan, J. The Fractographic Analysis of Tensile and Fatigue Fracture Surfaces in Secondary A356 Aluminum Alloy with a Higher Concentration of Iron. *Defect. Diffus. Forum* **2023**, *422*, 15–20. [CrossRef]
39. Zhong, Q.; Zhao, Z.; Zhang, Z. Development of "Fractography" and Research of Fracture Micromechanism. *Jixie Qiangdu* **2005**, *27*, 358–370.
40. Wang, Z.; Lin, X.; Kang, N.; Chen, J.; Tan, H.; Feng, Z.; Qin, Z.; Yang, H.; Huang, W. Laser powder bed fusion of high-strength Sc/Zr-modified Al–Mg alloy: Phase selection, microstructural/mechanical heterogeneity, and tensile deformation behavior. *J. Mater. Sci. Technol.* **2021**, *95*, 40–56. [CrossRef]
41. Cini, A.; Irving, P.E. Development of fatigue cracks from mechanically machined scratches on 2024-T351 aluminium alloy—Part I: Experimentation and fractographic analysis. *Fatigue Fract. Eng. Mater. Struct.* **2017**, *40*, 776–789. [CrossRef]
42. Zhong, Q.; Zhao, Z. *Fractography*, 1st ed.; Higher Education Press: Beijing, China, 2006; pp. 146–153.

metals

MDPI

Article

Ultrafine Grain 316L Stainless Steel Manufactured by Ball Milling and Spark Plasma Sintering: Consequences on the Corrosion Resistance in Chloride Media

Eric Hug [1,*], Clément Keller [2], Cendrine Folton [1], Jade Papin [1], Kostiantyn Tabalaiev [3] and Gaël Marnier [4]

[1] Laboratoire de Cristallographie et Sciences des Matériaux, University of Caen Normandie, Normandie Université, CNRS, 6 Bd Maréchal Juin, 14000 Caen, France
[2] Laboratoire Génie de Production, Université de Technologie de Tarbes Occitanie Pyrénées, Avenue d'Azereix, 65000 Tarbes, France
[3] CRMC, Industeel, 56 rue Clémenceau, 71202 Le Creusot, France
[4] ENSICAEN, 6 Bd Maréchal Juin, 14050 Caen, France
* Correspondence: eric.hug@ensicaen.fr

Abstract: This paper reports experimental results concerning the corrosion of 316L austenitic stainless steels produced by ball milling and spark plasma sintering in NaCl electrolyte. Specimens with grain sizes ranging from 0.3 μm to 3 μm, without crystallographic texture, were obtained and compared with a cast that is 110 μm in grain size and an annealed reference. The potentiodynamic experiments showed that the reduction in grain size leads to a degradation of the electrochemical passivation behavior. This detrimental effect can be overcome by appropriate passivation in a HNO_3 concentrated solution before consolidation. The Mott–Schottky measurements showed that the semiconducting properties of the passive layer do not vary significantly on the grain size, especially the donor density, which is responsible for the chemical passivation breakdown by chloride anions. The total electrical resistance of the layer, measured by impedance spectroscopy is always lower than the one of a cast and annealed 316L, but it slightly increases with a reduction in grain size in the ultrafine grain range. This is followed by a slight increase in the thickness of the oxide layer. The effect of chloride ions is very pronounced in terms of passivation breakdown if the powder is not passivated prior to sintering. This leads to the nucleation and growth of subsurface main pits and the formation of secondary satellite pits, especially for the smallest grain sizes. Passivation of the 316L powder before sintering has been found to be an effective way to prevent this phenomenon.

Keywords: corrosion; passivity; pitting; grain size; 316L; SPS; ball milling

1. Introduction

Stainless steels are very well-known alloys in which a certain amount of chromium (more than 11% in weight) is added to ordinary steels to improve corrosion resistance. Adding chromium to traditional steels also improves other properties, such as mechanical properties or surface characteristics, making them suitable for a wide range of applications. Within the family of stainless steels, AISI 316L exhibits attractive mechanical properties, as well as good formability and weldability, making it a good candidate for industrial applications in aggressive environments (nuclear [1], biomedical [2], marine [3], and petrochemical [4]). An austenitic structure generally ensures good corrosion resistance and helps us avoid microscopic galvanic corrosion products between residual martensite and austenite [5]. However, this austenitic microstructure, without strengthening components such as residual martensite, leads to suboptimal mechanical properties, which can be detrimental for some applications [6].

Reducing the grain size of fully austenitic 316L is one strategy for increasing the yield stress, but this is often accompanied by a decrease in ductility [7,8] due to the use of severe plastic deformation processes [9]. An alternative is to produce these alloys by powder metallurgy [10]. In this case, the materials are consolidated by solid-state sintering at a low temperature and pressure. However, mechanical and corrosion properties can be limited by the inherent porosity when compared to their wrought and cast counterparts. By using higher sintering temperatures [11] and controlling the sintering atmosphere [12], it has been found that the density of the material can be increased, resulting in improved properties. Manipulating the powder morphology and using flash-sintering techniques such as Spark Plasma Sintering (SPS) [13] are also efficient ways to control the porosity and to adjust the final properties of structural parts. This can be achieved by mechanical milling of the powder prior to consolidation [14]. The elaboration of Ultra-Fine Grain (UFG) specimens [15] with an average grain size from 0.1 μm to 1 μm, typically, can also be obtained in this way. The relative densities of alloys obtained by ball milling and SPS are typically between 90% and 99% [16,17], higher than those of conventional sintered parts. The difference in density is mainly due to the presence of porosity but also oxides resulting from the oxygen contamination during powder milling [18].

In previous works [19,20], it was shown by the authors that the mechanical properties of 316L stainless steel can be tuned in terms of strength and ductility by powder metallurgy, using ball milling and SPS. High-density fully austenitic and UFG samples can be produced by this method, with densities in the range of 93–99.6%. A higher yield stress was obtained compared to a traditional cast and annealed alloy, combined with a remarkable ductility, especially when bimodal powders were used [21,22]. In addition, SPS samples with a micrometric grain size exhibit an equivalent corrosion resistance in NaCl electrolyte compared to the traditional alloy [20]. The low density of porosity prevented the development of pitting mechanisms, as reported for conventionally sintered stainless steels [23].

Moreover, in a recent work [24], an improvement in the electrochemical passivity for the smallest grain sizes in neutral nonaggressive Na_2SO_4 electrolyte was characterized. In particular, the effect of pre-passivation of the ball-milled powder prior to sintering was clearly demonstrated. In order to provide a clear overview of the corrosion resistance of the promising UFG 316L manufactured by SPS, this paper is dedicated to the influence of a chloride environment, which is known to deeply affect the integrity of this material. Potentiodynamic experiments in NaCl electrolytes were carried out to analyze the overall corrosion behavior. The physical characteristics of the passive layer were studied using an impedance spectroscopy analysis. The effect of grain size and grain-boundary nature on the stability of the passive layer in chloride media is discussed in this paper, supported by microstructural observations of the pitted surfaces after the potentiodynamic experiments.

2. Experimental Details

2.1. Sintering Elaboration and Microstructure Observations

316L stainless steel samples were elaborated by spark plasma sintering using a precursor commercial atomized powder. The methodology and main experimental parameters are displayed in Figure 1. The first series was sintered by SPS (FCT System GmbH, HD25 SPS apparatus, Effelder-Rauenstein, Germany), using the as-received commercial powder (CP samples). Based on our previous investigations, optimal sintering conditions (pressure, temperature, and dwell time) were employed [20], leading to relative densities higher than 99%, and an average grain size of 2–3 μm.

Ball milling (Fritsh planetary pulverisette 7, Idar-Oberstein, Germany) of the as-received powder was carried out before sintering to decrease the final grain size, following a methodology previously optimized [19]. Using this previous work, a second series were built (A1, A2, and A3), with grain size ranging from 0.4 μm to 1 μm and a relative density of about 95%.

A third series of samples, called B1, B2, and B3, were considered in this work, consisting of immerging and then ball-milled powder in a HNO_3 acid solution before SPS consolidation [25]. Finally, a cast and annealed (CA) alloy, 110 μm in grain size, was also used for comparison.

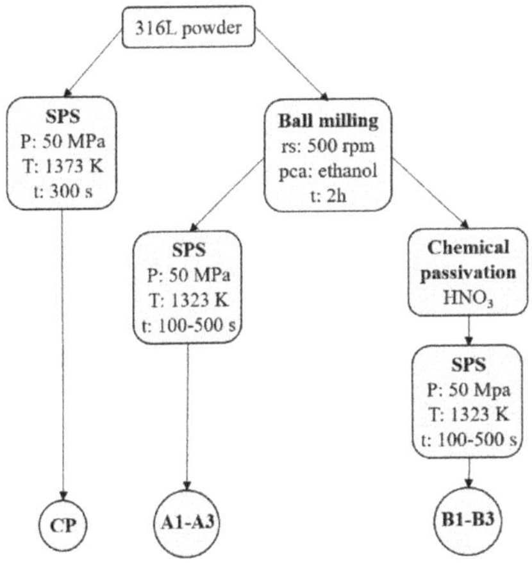

Figure 1. Flow diagram of the elaboration methodology, experimental details concerning ball milling (rs, rotation speed; pca, process control agent; and t, milling time) and SPS (P, pressure; T, temperature; and t, dwell time), and corresponding 316L samples.

Densities were obtained by a traditional Archimedes method, and the microstructure was studied by Electron Back-Scattered Diffraction (EBSD) and Scanning Electron Microscopy (SEM). The EBSD acquisition system used was a QUANTAX EBSD (Bruker, Berlin, Germany) mounted on a Zeiss Supra 55 (Jena, Germany). Observations of the corroded surfaces after electrochemical experiments were performed by light microscopy and laser confocal microscopy. Pitting observations were performed after potentiodynamic polarization experiments, which were stopped at a current density of 10^{-1} mA/cm^2 in the anodic region.

2.2. *Electrochemical Experiments*

Potentiodynamic polarization, $J(E)$, curves (J, current density; and E, potential) were obtained using a three-electrode electrochemical cell and a potentiostat workstation (SP-300 Biologic, Grenoble, France). The potential reference electrode was a Ag/AgCl one, saturated with AgCl-KCl electrolyte. A platinum grid was employed as a counter electrode. Five experiments were performed for each condition to confirm the good repeatability of the results. The tests were carried out in aerated NaCl 0.85 M (3 wt.%) at room temperature. Prior to all experiments, the working electrode surface was mechanically polished, and then a cathodic reduction was applied to reduce surface oxides. Samples were immerged in NaCl in open-circuit conditions for 1 h, and then a potentiodynamic test was performed with a voltage rate of 0.22 mV/s.

Electrochemical Impedance Spectroscopy (EIS) and Mott–Schottky (MS) were performed in open-circuit conditions to obtain information on the electrical and semiconducting properties of the passive layer. For EIS, 10 mV in potential amplitude was imposed, with the frequency decreasing from 10 kHz up to 1 mHz. Capacitance experiments were conducted to obtain MS graphs. They were carried out at 1 kHz and 10 mV in potential amplitude, from open-circuit potential to −1 V/Ag-AgCl. As for the potentiodynamic tests, five experiments were carried out for each configuration.

3. Results and Discussion

3.1. *Brief Microstructural Overview of the Samples*

Details concerning microstructural features of the samples manufactured by ball milling and SPS are given in previous publications [19,20,22,24], and are only briefly reviewed here. Typical microstructures analyzed by EBSD are displayed in Figure 2, whereas Table 1 summarizes a few microstructural characteristics of the alloys.

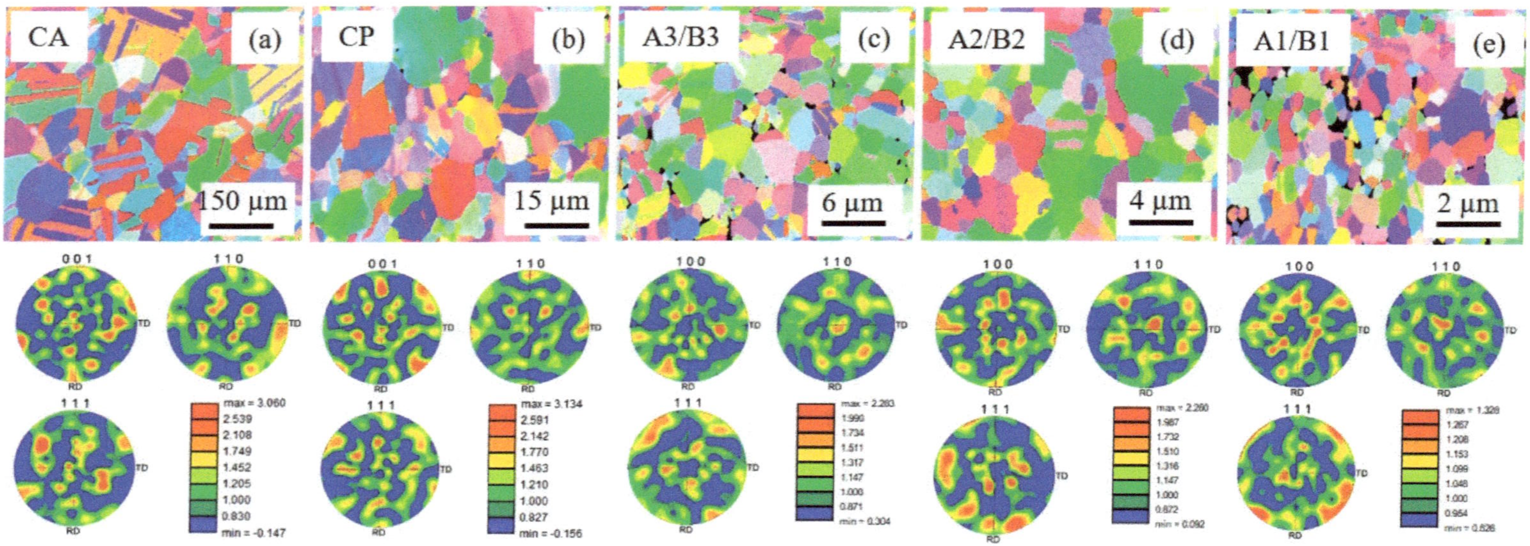

Figure 2. Grain orientation maps and associated pole figures for the 316L samples: (**a**) CA, (**b**) CP, (**c**) A3/B3, (**d**) A2/B2, and (**e**) A1/B1. Pre-passivated (B) and not passivated (A) samples exhibit the same microstructural features.

Table 1. Some microstructural and mechanical characteristics of 316L specimens (following data coming from [19,20,22,24]). *d* is the average grain diameter, ρ the relative density, LAGB is the low-angle grain boundary amount, HAGB is the high-angle grain boundary amount, Σ_3 is the twin amount, and MPD is the Maximal Pole Density of the crystallographic textures.

Sample Code	ρ (%)	*d* (µm)	Σ_3 (%)	LAGB (%)	HAGB (%)	MPD (MRD)
CA	100	110 ± 20	34.6	28.9	71.1	3.35
CP	99.6	2.7 ± 1.8	28.7	23.2	76.8	3.13
A1, B1	95.3	0.33 ± 0.09	19.1	13.2	86.8	1.33
A2, B2	95.1	0.67 ± 0.07	28.4	18.8	81.2	2.28
A3, B3	94.7	0.88 ± 0.08	29.7	10.2	89.8	2.28

The combination of ball milling and SPS is an efficient way to reduce the grain size of 316L without residual stresses [20]. The main difference between the samples is the final grain size, which is distributed over three orders of magnitude, in the range [100 nm–100 µm] (Figure 3). B specimens exhibit the same density and the same grain sizes and LAGB/HAGB ratio as A samples.

The average density is higher than 95%, and no crystallographic texture is depicted, independently on the manufacturing conditions (maximum pole density always less than 5 MRD and no preferential orientations emerging on the pole figures). The slightly reduced density of specimens manufactured by ball milling is mainly due to the oxides and porosities present in a few of the sintered samples. The maximum oxide fraction is found to be around 10% [19]. The quantities of the low-angle grain boundary (LAGB) and Σ_3 twin boundary fluctuate around 20%, without specific tendency. Finally, a fully austenitic phase is always present, without any presence of martensite (confirmed by XRD analysis, but not given in this publication for the sake of brevity).

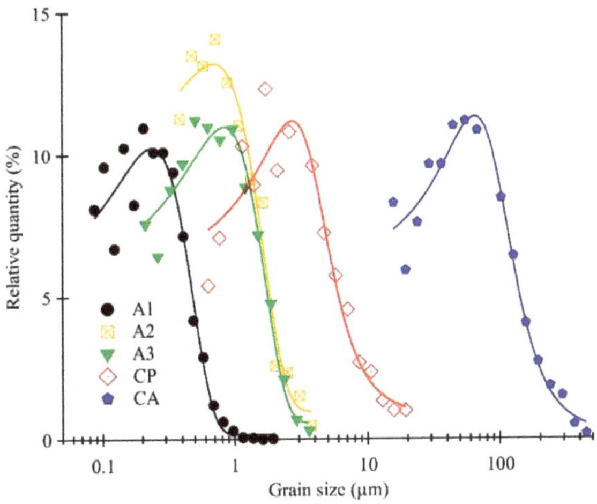

Figure 3. Grain size distributions corresponding to 316L samples shown in Figure 2.

3.2. Corrosion Behavior

Figure 4 shows examples of *J(E)* curves obtained for samples with different grain sizes. The impact of the passivation of the ball-milled powder before consolidation is clearly evidenced. Corrosion parameters extracted from *J(E)* curves are compiled in Table 2.

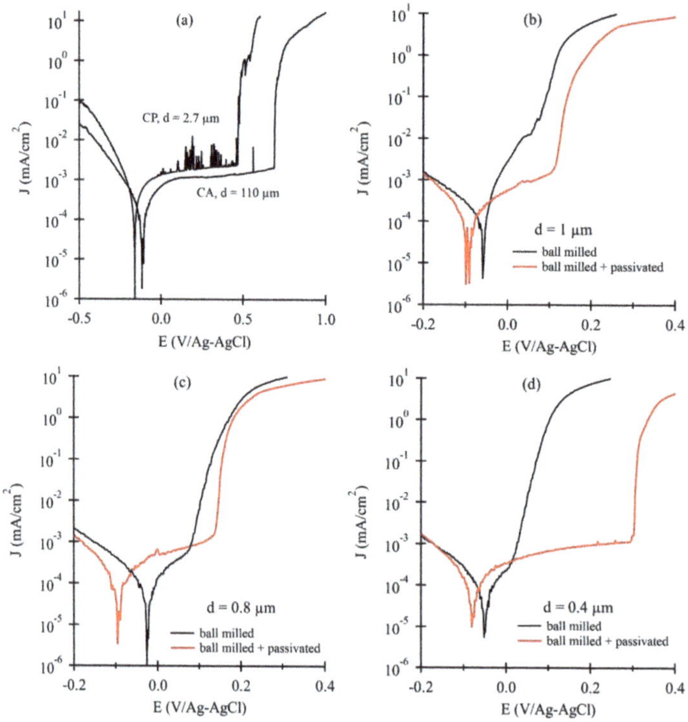

Figure 4. (**a**) Potentiodynamic curves of CA and CP 316L. Impact of ball milling and passivation of the powder for three final grain sizes: (**b**) d = 1 µm (A3 and B3 samples), (**c**) d = 0.8 µm (A2 and B2 samples), and (**d**) d = 0.4 µm (A1 and B1 samples).

Table 2. Mean values of corrosion parameters from *J(E)* experiments performed on 316L samples. Potentials are given versus Ag-AgCl electrode.

Sample Code	E_{cor} (mV)	J_{cor} * (µA/cm²)	J_{pas} (µA/cm²)	E_{pit} (mV)	L (mV)
CA	−122		0.93	628	750 ± 175
CP	−136		1.71	267	416 ± 154
A1	−52	0.85	---------	-----	-----
A2	−25	0.95	---------	-----	-----
A3	−59	0.77	---------	-----	-----
B1	−82		0.65	296	438 ± 137
B2	−97		0.57	137	234 ± 51
B3	−91		0.61	105	196 ± 38

* J_{cor} obtained from Tafel plots, only for A samples, when no passivity is depicted.

The impact of the powder preparation before SPS consolidation on the overall electrochemical behavior is very pronounced. A large passivity domain is depicted in CA samples, in a potential range of $L = E_{pit} - E_{cor} = 750$ mV (Figure 4a), with E_{pit} representing the pitting potential. CP samples exhibit a shorter and very instable passive potential plateau with the existence of numerous vertical peaks revealing metastable pit nucleation, followed by repassivation of the pits (Figure 4a). The passivation current density, J_{pas}, is also higher for CP than for the traditional CA specimen, suggesting the degradation of the corrosion resistance in the passive region of SPS samples directly consolidated with the commercial powder.

Pitting mechanisms due to chloride anions tend to strongly shorten the length of the passive layer for A-type UFG specimens, leading to a premature breakdown of the passivity. E_{pit} decreases when the grain size decreases, and the passive plateau even vanishes for A samples (Figure 4b–d). The current density increases in the pseudo passive plateau, attributed to crevice corrosion mechanisms starting from open porosities [25]. A well-defined passive plateau is, however, again depicted for B samples in the figures, revealing the effect of pre-passivation of the ball-milled powder before SPS. J_{pas} is smaller for B samples than for CA cast ones, suggesting the presence of a more efficient oxide layer, even if J_{pas} slightly increases with a decrease in grain size. The pre-passivation of the powder also increases the stability of the passive layer for the smallest grain diameters (Figure 4d). Indeed, the length of the passivation plateau, L, is about 438 mV for B1 specimen (d = 400 nm), which is the best value obtained for SPS samples, and grossly half the one measured on cast and annealed samples (L = 750 mV in average). Hence, taken into account the large scattering of the values of L_p (see values in Table 2), due to the stochastic character of the pitting nucleation in stainless steels [26,27], the combination of ball milling and passivation of the powder is an efficient way to obtain 316L samples with improved corrosion resistance in a chloride environment, close and even slightly better than the one of cast and annealed material.

3.3. Electrical and Semiconducting Characteristics of the Surface Oxide

Typical Nyquist plots are displayed in Figure 5 for some representative samples. An open depressed capacitive loop is always evidenced independently on the elaboration conditions, with a higher radius of the loop representing a better corrosion resistance [28].

Samples obtained by powder metallurgy exhibit a capacitive loop with a diameter smaller than the one obtained for the cast and annealed stainless steel. Unlike the results obtained for Na_2SO_4 [24], the presence of chloride ions seems then to degrade the electrical properties of the oxide layer, and this effect is more pronounced for specimens sintered from ball-milled powders. No significant difference is, moreover, depicted between the A and B samples, independent of the final grain size.

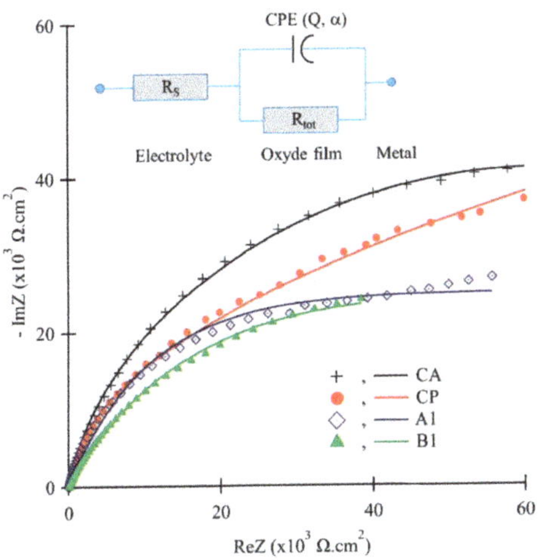

Figure 5. Examples of Nyquist diagrams for 316L specimens with various grain sizes. An equivalent electrical circuit is given in the inset.

Despite the double-layer character of the passive film on 316L surface [29–31], a simpler and sufficient model was adopted in this paper by considering the total resistive and capacitive answer of the oxide layer, as frequently used by other researchers [32–34]. The equivalent electrical circuit (inset in Figure 5) consists in the electrolyte resistance, R_s, and an (R_{tot}, CPE) element. The R_{tot} is the total polarization resistance of the interface. A Constant Phase Element (CPE) was chosen to represent the impedance of the oxide layer. The impedance of the CPE element is as follows:

$$Z_{CPE} = \frac{1}{Q(2\pi f j)^{\alpha}} \quad (1)$$

where f is the frequency, Q is the CPE, α is the capacitor coefficient, and j is the imaginary number.

The experimental results are well represented by this simple electrical model, as shown in Figure 5 by the solid lines, especially for frequencies higher than a few hundred Hertz, representative to the internal oxide layer. The model parameters, given the best fit for all the samples, are resumed in Table 3. α is almost constant for all samples, with an average value $\alpha = 0.89 \pm 0.02$; a value close to $\alpha = 1$ is representative of a pure capacitive behavior. No clear tendency is observed concerning Q, which takes values between 15 $\mu F \cdot s^{\alpha-1}/cm^2$ and 30 $\mu F \cdot s^{\alpha-1}/cm^2$.

Table 3. Impedance-model parameters corresponding to the electrical equivalent circuit represented in Figure 5 for 316L samples in NaCl electrolyte.

Sample Code	R_s ($\Omega \cdot cm^2$)	R_{tot} ($k\Omega \cdot cm^2$)	Q ($\mu F \cdot s^{\alpha-1}/cm^2$)	α	C_{eff} ($\mu F/cm^2$)	δ (nm)
CA	22.9	201.28	16.1	0.90	31.0	2.6
CP	15.1	134.26	28.4	0.89	60.1	1.3
A1	17.3	225.69	14.6	0.89	28.9	2.8
A2	21.7	199.23	18.7	0.91	33.9	2.3
A3	18.6	175.68	26.0	0.87	65.5	1.2
B1	16.5	256.26	18.3	0.90	34.5	2.3
B2	22.4	218.98	14.7	0.87	49.1	2.4
B3	19.2	204.01	21.2	0.88	48.1	1.7

When using the single CPE model adopted in this work, an effective capacity, C_{eff}, of the surface layer can be estimated by using the following relation [32]:

$$C_{eff} = Q^{\frac{1}{\alpha}} \cdot \left(\frac{R_S R_{tot}}{R_S + R_{tot}} \right)^{\frac{1-\alpha}{\alpha}} \cong Q^{\frac{1}{\alpha}} \cdot R_S^{\frac{1-\alpha}{\alpha}} \quad (2)$$

The approximation in the right part of Equation (2) is supported by the fact that $R_{tot} >> R_s$ (see Table 3). Then, considering the oxide surface layer as an ideal plane capacitor, its thickness, δ, can be computed by using the following well-known formulae:

$$\delta = \frac{\varepsilon_0 \varepsilon_r}{C_{eff}} \quad (3)$$

$\varepsilon_r = 15.6$ [35] is the relative dielectric constant of oxide film in 316L. $\varepsilon_0 = 8.85 \times 10^{-14}$ F/cm is the vacuum permittivity. The values of C_{eff} and δ are reported in Table 3. δ is around a few nanometers, in accordance with previous results [32], and it does not vary with the grain size. However, the larger values (higher than 2 nm) are found for the cast and annealed sample, as well as for the smallest grain sizes. The CP, A3, and B3 specimens present the smallest oxide thickness, explaining, in part, that these samples are weakly resistant in corrosive NaCl electrolyte (Figure 3). Nanostructuration increases the oxide thickness when the smallest grain sizes are reached (A1 and B1 samples), given similar values to cast and annealed ones, independently of the chemical passivation step of the powder.

The polarization resistance, R_{tot}, sensitively depends on the grain size, d (Figure 6), following a linear relation with $1/d$. The same evolution was reported in a Na_2SO_4 electrolyte [24]. However, it can be noted that B samples do not exhibit the same behavior as A samples. R_{tot} values are higher for B samples but tend toward the same values as those depicted in A samples for the smallest grain diameters.

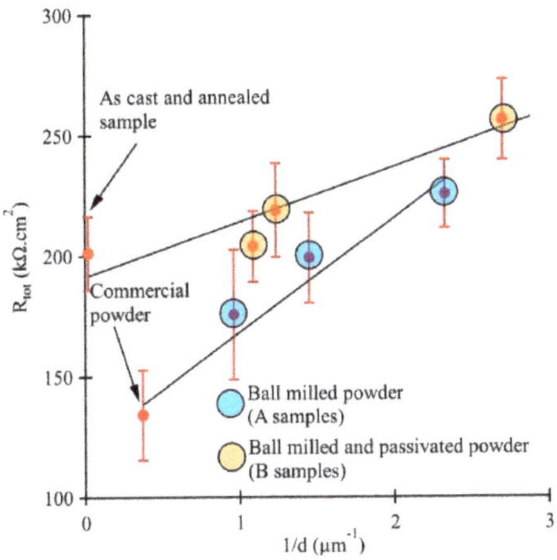

Figure 6. Evolution of the polarization resistance, R_{tot}, with $1/d$ for all the 316L samples.

The evolution of $1/C^2$ with E in the cathodic potential region is displayed in Figure 7 (Mott–Schottky (MS) plots). These diagrams display two linear stages delimited above by the flat band potential, E_{FB} = −0.55 V/Ag-AgCl, not dependent on the elaboration conditions. The negative slope corresponds to a p-type semiconductor and the positive slope is related to a n-type oxide. A duplex oxide layer is therefore evidenced, with an inner Cr_2O_3 p-type layer and an outer Fe_2O_3 n-type oxide [33]. The two linear relationships between $1/C^2$ and E can be written as follows [29]:

$$\frac{1}{C^2} = \left(\frac{2}{\varepsilon_0\varepsilon_r e N_D}\right)\left(E - E_{FB} - \frac{kT}{e}\right) \quad (4)$$

$$\frac{1}{C^2} = \left(\frac{2}{\varepsilon_0\varepsilon_r e N_A}\right)\left(-E + E_{FB} - \frac{kT}{e}\right) \quad (5)$$

where N_D is the charge carrier donor density, and N_A is the charge carrier acceptor density. $e = 1.6 \times 10^{-19}$ C is the elementary electrical charge, $k = 1.38\text{x}10^{-23}$ J/K is the Boltzmann constant, and T = 293 K is the room temperature. $kT/e \approx 0.025$ V and is negligible in the potential range of interest of Figure 7.

Table 4 sums up the representative values N_A, N_D, and E_{FB} for each 316L specimen. A good agreement is found with the previous literature [30,35,36]. N_A and N_D are in the order of 10^{21} cm^{-3}, representing highly doped semiconductors.

The total carrier density, $N = N_A + N_D$, strongly increases with the grain refinement, as displayed in the inset of Figure 7. The electron transfer through the oxide surface is then enhanced because the film is more defective. The stability of the oxide layer is less effective for the smallest grain sizes [37]. Indeed, Cl^- anions can be more easily transferred through the thickness of the oxide layer, filling the O^{2-} vacancies, and this can ultimately cause the oxide layer to fracture.

Prepassivation of the powder has, however, a beneficial effect on this deleterious mechanism by increasing R_{tot} (Figure 6), and it slightly increases the oxide thickness, δ, when the grain size is reduced (Table 3). The overall passivation current density is then smaller for the B samples, than for the CA, CP, or A specimens (Figure 4). However, the pitting potential is always smaller for SPS samples, suggesting a greater instability of the passive layer. Prepassivation combined with nanostructuration enhances, however, the pitting resistance of SPS specimens. More detailed observations of pits are given in the next part of the paper in order to better understand the underlying pitting mechanisms.

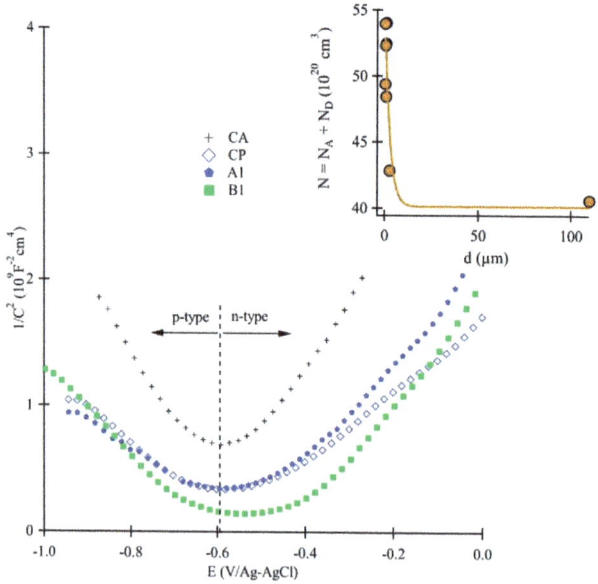

Figure 7. Mott–Schottky plots around the flat band potential for passive films formed on some 316L specimens. The variation of $N = N_A + N_D$, along with the grain diameter, d, is represented in inset.

Table 4. Values of N_A, N_D, and E_{FB} for all the 316L specimens. (Potential is given versus Ag-AgCl electrode).

Sample Code	N_A (10^{20} cm^3)	N_D (10^{20} cm^3)	E_{FB} (mV)
CA	15.22	25.39	−590
CP	20.14	22.70	−556
A1	31.45	22.50	−557
A2	29.70	22.62	−556
A3	32.77	21.25	−568
B1	27.46	21.92	−543
B2	28.50	19.91	−571
B3	29.57	22.88	−566

3.4. Role of the Prepassivation of the Ball-Milled Powder to Prevent Pitting Mechanisms

In the presence of Cl^- anions, the passive film grown on surface samples becomes sensitive to chemically induced breakdown, leading to pitting mechanisms. Typical corroded surfaces and pitting behavior after potentiodynamic tests in NaCl electrolyte ($J_{max} = 10^{-1}$ mA/cm^2) are displayed in Figure 8 for the A and B specimens.

Figure 8 shows that pitting is strongly correlated to the average grain diameter for A samples. A1 samples (d = 0.4 μm) exhibit a very important pit density, leading to a generalized corrosion of the exposed area (Figure 8a). A2 samples (d = 0.75 μm) are subjected to localized corrosion phenomena, with numerous pits on the exposed surface

(Figure 8b), but a large part of the sample remains free of pits at this observation scale. Concerning the B1 and B2 samples (Figure 8c,d), obtained with the passivated powder, no pitting is depicted on the exposed area at this observation level. For these samples, the breakdown of passivity observed in potentiodynamic curves in Figure 4d is, therefore, mainly produced by the smallest pits, which are mainly nucleated on grain boundaries and porosities [38].

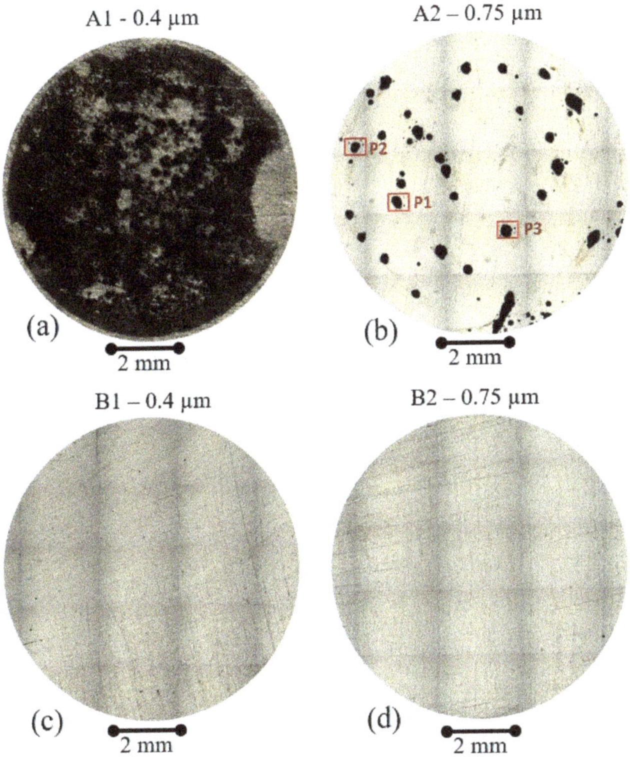

Figure 8. Corrosion landscape of SPS 316L with ball-milled (A samples) powder and ball-milled and passivated powder (B samples): (**a**) A1, (**b**) A2, (**c**) B1, and (**d**) B2. Average grain sizes are given for each sample.

From a material-health point of view, the localized corrosion observed on A2 is more harmful than the almost-generalized corrosion observed on A1. Indeed, in the first case, where the attacked surface being smaller than the surface exposed to the electrolyte, the local current density is much greater than $J_{max} = 10^{-1}$ mA/cm^2. Since the corrosion rate is directly proportional to J, due to Faraday's law, A2 will crumble much faster than A1.

Looking in more detail at the geometric features of the pit in A2 specimens, we see that the ratio between the depth, P_{max}, and the equivalent diameter, D_{eq}, of some representative pits, in function of D_{eq}, is displayed in Figure 9. Figure 10 also illustrates some examples of pits depicted in A2 surfaces after potentiodynamic tests. Large flat-bottomed main pits can be observed, with a P_{max}/D_{eq} ratio lower than one, corresponding to filled square points in Figure 9. Around these large pits, smaller satellite pits are also observed, distributed around the main pit mouth. The P_{max}/D_{eq} ratio for these satellites can be higher (empty red square, Figure 9) or lower than one (empty blue square, Figure 9), depending on their distribution around the principal pit.

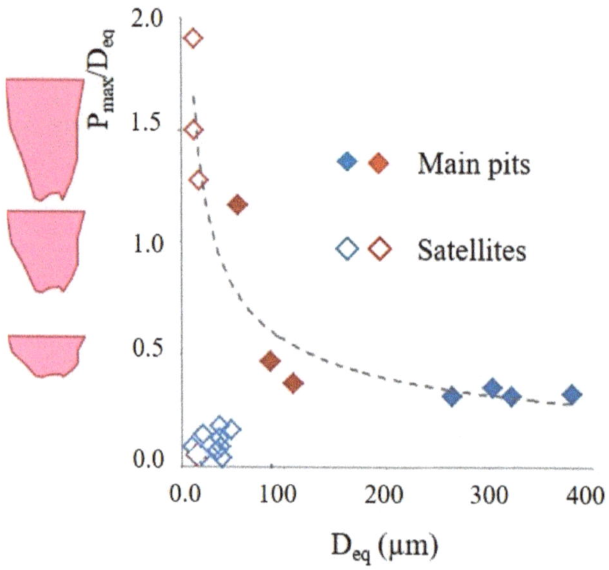

Figure 9. Evolution of the ratio between the maximal depth, P_{max}, and the equivalent diameter, D_{eq}, of several representative pits depicted in A2 samples. Schematic morphology of the pits in pink, depending on the ratio P_{max}/D_{eq}.

Under specific electrochemical conditions, pits tend to grow under the free surface of stainless steels. This leads to a so-called lacy cover on the pits [39]. This cover hinders the diffusion of corrosive and aggressive species, trapped in the bottom of the pit. Repassivation of such a pit is therefore prevented, and the corrosion propagates below the outer free surface [40]. The holes are produced by metal dissolution inside the pit, beneath its rim, undermining the metal surface and finally emerging outside, through the free surface [41]. When such satellite emergence occurs, this feature leads to the local uplift of the free surface, as observed by confocal imaging (Figure 11).

This mechanism is also supported by statistical measurements on all pits on the A2 surface, performed after successive polishings P1, P2, . . ., P8, in order to access to a pseudo 3D cartography of the main pit and associated satellites (Figure 12). The average diameter of pits after polishing tends to increase up to a depth of about 40 µm, and then the diameter reaches a plateau and finally decreases for higher depths (Figure 12a). This is also illustrated in Figure 12b, which displays the successive height profiles of a given pit, and in Figure 12c, representing the iso-contour of a pit after successive polishings. Corrosion under the surface of the steel is evidenced both through these two examples and statistically for all the pits on the sample surface (Figure 12a), and it can be schematically drawn, as shown in Figure 12d.

Using HNO_3 acid solution to passivate the ball-milled powder before SPS prevents this pitting growth mechanism, leaving the exposed surface to the NaCl electrolyte free from large pits (B samples, Figure 8c,d) and, therefore, making the material more resistant.

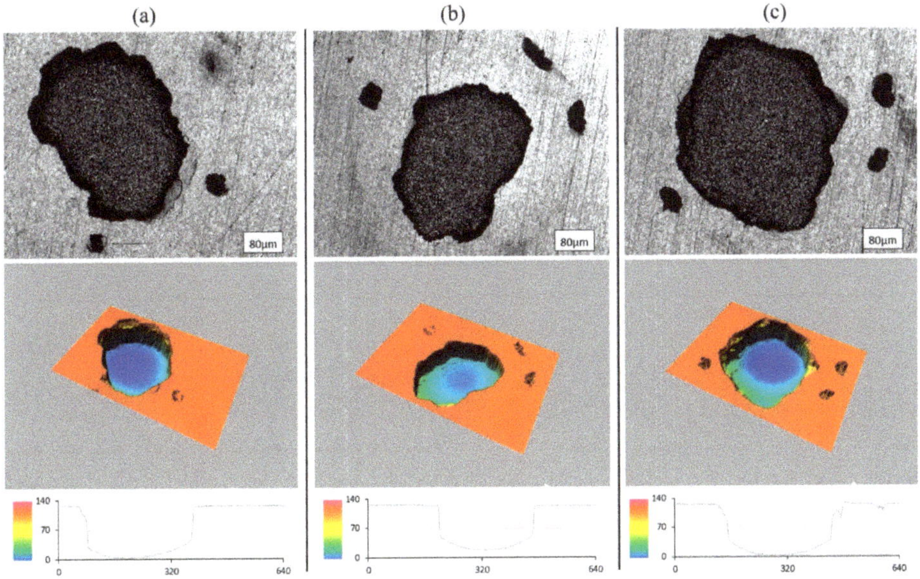

Figure 10. Typical pits observed on A2 surfaces after exposition to NaCl environment during potentiodynamic experiment (J_{max} = 10^{-1} mA/cm^2). (**a–c**) Pits P1, P2, and P3, highlighted by red squares in Figure 8b. For each pit (vertical images) are given, from up to bottom, the image by light microscopy, the 3D profile by laser confocal microscopy, and a depth profile (values are in µm).

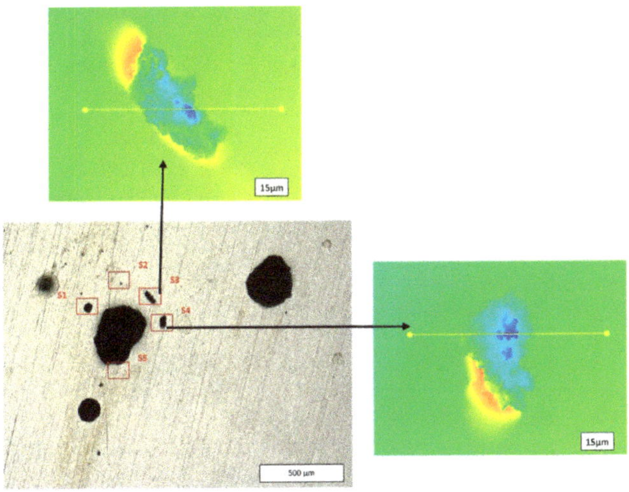

Figure 11. P2 pit (cf. Figure 8b) depicted on A2 surfaces, with five smaller satellite pits around. Uplift of the surface matter around the satellites is evidenced as red areas depicted by confocal 2D images.

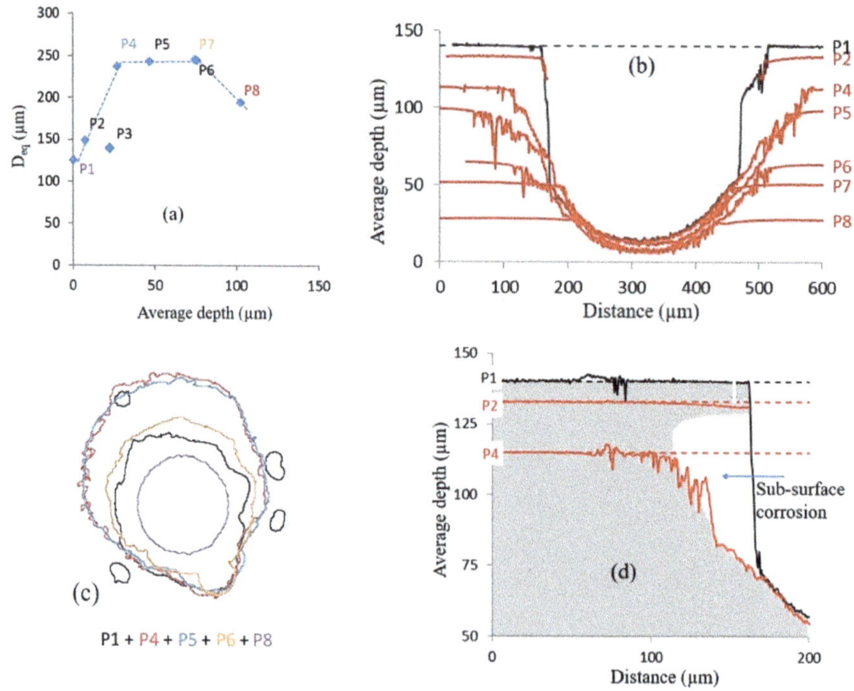

Figure 12. Evolution of the pit morphology with several successive polishings, P1, P2, . . ., P8. (**a**) Variation of the average equivalent diameter with average depth. Statistical measurements performed on all the pits present in A2 surface. (**b**) Evolution of the profile along the yellow line of a pit shown in inset. (**c**) Iso-contour of the diameter of a pit and its satellites (black) after several polishing levels. (**d**) Schema of the development of subsurface corrosion for a main pit.

4. Conclusions

In this work, the corrosion resistance in the NaCl environment of UFG 316L steels produced by ball milling and SPS was studied. The main conclusions of this work are as follows:

- The passivity is improved when the powder is passivated in HNO_3 prior to sintering, especially for the smallest grain sizes.
- Reducing the grain diameter in the UFG range increases the electrical resistance of the oxide film.
- The total charge carrier density of the oxide layer decreases as the grain size increases. Cl^- anions are then less able to transfer through the thickness of the outer layer, and then the chemical breakdown of the passive layer is more difficult to achieve.
- The mechanism of pitting consists of nucleation and subsurface growth of the main pits, followed by the formation of satellite pits distributed around the main pits.
- The pre-passivation of the powder prevents this pitting nucleation and growth mechanism. As a result, the samples with the smallest grain diameters made with the pre-passivated powder develop a better resistance to pitting than those made with the commercial powder, as well as improved corrosion properties in chloride environments.

Author Contributions: Conceptualization, E.H.; methodology, C.K. and K.T.; validation, C.K., C.F. and G.M.; formal analysis, E.H. and C.F.; investigation, C.F., J.P., K.T. and G.M.; data curation, J.P.; writing—original draft, E.H.; writing—review and editing, E.H., C.K., C.F. and J.P.; supervision,

E.H.; project administration, C.K. All authors have read and agreed to the published version of the manuscript.

Funding: This research was funded by the financial support of the French Agence Nationale de la Recherche (ANR), through the program "Investissements d'Avenir" (CMAIN Project, ANR-10-LABX-09-01), LabExEMC3.

Data Availability Statement: The data are not publicly available due to industrial confidential restrictions.

Conflicts of Interest: Author Kostiantyn Tabalaiev was employed by the CRMC, Industeel. Author Gaël Marnier was employed by the ENSICAEN. The remaining authors declare that the research was conducted in the absence of any commercial or financial relationships that could be construed as a potential conflict of interest.

References

1. He, Q.; Pan, F.; Wang, D.; Liu, H.; Guo, F.; Wang, Z.; Ma, Y. Microstructure and properties of 316L stainless steel foils for pressure sensor of pressurized water reactor. *Nucl. Eng. Technol.* **2021**, *53*, 172–177. [CrossRef]
2. Zhang, E.; Zhao, X.; Hu, J.; Wang, R.; Fu, S.; Qin, G. Antibacterial metals and alloys for potential biomedical implants. *Bioact. Mater.* **2021**, *6*, 2569–2612. [CrossRef] [PubMed]
3. Jena, G.; Anandkumar, B.; Sofia, S.; George, R.P.; Philip, J. Fabrication of silanized GO hybrid coating on 316L SS with enhanced corrosion resistance and antibacterial properties for marine applications. *Surf. Coat. Technol.* **2020**, *402*, 126295. [CrossRef]
4. D'Andrea, D. Additive Manufacturing of AISI 316L Stainless Steel: A Review. *Metals* **2023**, *13*, 1370. [CrossRef]
5. Liu, E.; Zhang, Y.; Zhu, L.; Zeng, Z.; Gao, R. Effect of strain-induced martensite on the tribocorrosion of AISI 316L austenitic stainless steel in seawater. *RSC Adv.* **2017**, *7*, 44923–44932. [CrossRef]
6. Song, R.-B.; Xiang, J.-Y.; Hou, D.-P. Characteristics of Mechanical Properties and Microstructure for 316L Austenitic Stainless Steel. *J. Iron Steel Res. Int.* **2011**, *18*, 53–59. [CrossRef]
7. Krasilnikov, N.; Lojkowski, W.; Pakiela, Z.; Valiev, R. Tensile strength and ductility of ultra-fine-grained nickel processed by severe plastic deformation. *Mater. Sci. Eng. A* **2005**, *397*, 330–337. [CrossRef]
8. Sun, L.G.; Wu, G.; Wang, Q.; Lu, J. Nanostructural metallic materials: Structures and mechanical properties. *Mater. Today* **2020**, *38*, 114–135. [CrossRef]
9. Valiev, R.Z.; Estrin, Y.; Horita, Z.; Langdon, T.G.; Zehetbauer, M.J.; Zhu, Y.T. Fundamentals of Superior Properties in Bulk NanoSPD Materials. *Mater. Res. Lett.* **2016**, *4*, 1–21. [CrossRef]
10. Choi, J.-P.; Lee, G.-Y.; Song, J.-I.; Lee, W.-S.; Lee, J.-S. Sintering behavior of 316L stainless steel micro–nanopowder compact fabricated by powder injection molding. *Powder Technol.* **2015**, *279*, 196–202. [CrossRef]
11. Pandya, S.; Ramakrishna, K.S.; Annamalai, A.R.; Upadhyaya, A. Effect of sintering temperature on the mechanical and electrochemical properties of austenitic stainless steel. *Mater. Sci. Eng. A* **2012**, *556*, 271–277. [CrossRef]
12. Kurgan, N. Effects of sintering atmosphere on microstructure and mechanical property of sintered powder metallurgy 316L stainless steel. *Mater. Des.* **2013**, *52*, 995–998. [CrossRef]
13. Preston, A.D.; Ma, K. Insight into the effects of pore size and distribution on mechanical properties of austenite stainless steels. *J. Mater. Sci.* **2021**, *56*, 17278–17295. [CrossRef]
14. Oleszak, D.; Grabias, A.; Pekala, M.; Swiderska-Sroda, A.; Kulik, T. Evolution of structure in austenitic steel powders during ball milling and subsequent sintering. *J. Alloy Compd.* **2007**, *434–435*, 340–343. [CrossRef]
15. Trzaska, Z.; Cours, R.; Monchoux, J.-P. Densification of Ni and TiAl by SPS: Kinetics and Microscopic Mechanisms. *Metall. Mater. Trans. A* **2018**, *49*, 4849–4859. [CrossRef]
16. Hu, Z.-Y.; Zhang, Z.-H.; Cheng, X.-W.; Wang, F.-C.; Zhang, Y.-F.; Li, S.-L. A review of multi-physical fields induced phenomena and effects in spark plasma sintering: Fundamentals and applications. *Mater. Des.* **2020**, *191*, 108662. [CrossRef]
17. Orrù, R.; Licheri, R.; Locci, A.M.; Cincotti, A.; Cao, G. Consolidation/synthesis of materials by electric current activated/assisted sintering. *Mater. Sci. Eng. R Rep.* **2009**, *63*, 127–287. [CrossRef]
18. Cipolloni, G.; Pellizzari, M.; Molinari, A.; Hebda, M.; Zadra, M. Contamination during the high-energy milling of atomized copper powder and its effects on spark plasma sintering. *Powder Technol.* **2015**, *275*, 51–59. [CrossRef]
19. Keller, C.; Tabalaiev, K.; Marnier, G.; Noudem, J.; Sauvage, X.; Hug, E. Influence of spark plasma sintering conditions on the sintering and functional properties of an ultra-fine grained 316L stainless steel obtained from ball-milled powder. *Mater. Sci. Eng. A* **2016**, *665*, 125–134. [CrossRef]
20. Marnier, G.; Keller, C.; Noudem, J.; Hug, E. Functional properties of a spark plasma sintered ultrafine-grained 316L steel. *Mater. Des.* **2014**, *63*, 633–640. [CrossRef]
21. Flipon, B.; de la Cruz, L.G.; Hug, E.; Keller, C.; Barbe, F. Elaboration of austenitic stainless steel samples with bimodal grain size distributions and investigation of their mechanical behavior. *AIP Conf. Proc.* **2017**, *1896*, 200007.
22. Flipon, B.; Keller, C.; Cruz, L.G.D.L.; Hug, E.; Barbe, F. Tensile properties of spark plasma sintered AISI 316L stainless steel with unimodal and bimodal grain size distributions. *Mater. Sci. Eng. A* **2018**, *729*, 249–256. [CrossRef]

23. García, C.; Martín, F.; de Tiedra, P.; Cambronero, L.G. Pitting corrosion behaviour of PM austenitic stainless steels sintered in nitrogen–hydrogen atmosphere. *Corros. Sci.* **2007**, *49*, 1718–1736. [CrossRef]
24. Hug, E.; Keller, C. Passivation properties of ultrafine grain 316L stainless steels manufactured by ball milling and spark plasma sintering. *Mater. Chem. Phys.* **2023**, *309*, 128398. [CrossRef]
25. Kargar, B.S.; Moayed, M.H.; Babakhani, A.; Davoodi, A. Improving the corrosion behaviour of powder metallurgical 316L alloy by prepassivation in 20% nitric acid. *Corros. Sci.* **2011**, *53*, 135–146. [CrossRef]
26. Liu, L.; Li, Y.; Wang, F. Pitting mechanism on an austenite stainless steel nanocrystalline coating investigated by electrochemical noise and in-situ AFM analysis. *Electroch. Acta* **2008**, *54*, 768–780. [CrossRef]
27. Racot, A.; Aubert, I.; Touzet, M.; Thiebaut, S.; Demesy, M. Statistical analysis of the pitting corrosion induced by potentiostatic pulse tests of wrought and SLM 316L stainless steels. *Corros. Sci.* **2022**, *197*, 110036. [CrossRef]
28. Orazem, M.E.; Tribollet, B. *Electrochemical Impedance Spectroscopy*; Wiley: Hoboken, NJ, USA, 2008.
29. Gomes, W.P.; Vanmaekelbergh, D. Impedance spectroscopy at semiconductor electrodes: Review and recent developments. *Electroch. Acta* **1996**, *41*, 967–973. [CrossRef]
30. Trisnanto, S.R.; Wang, X.; Brochu, M.; Omanovic, S. Effects of crystallographic orientation on the corrosion behavior of stainless steel 316L manufactured by laser powder bed fusion. *Corros. Sci.* **2022**, *196*, 110009. [CrossRef]
31. Wang, Z.; Paschalidou, E.-M.; Seyeux, A.; Zanna, S.; Maurice, V.; Marcus, P. Mechanisms of Cr and Mo Enrichments in the Passive Oxide Film on 316L Austenitic Stainless Steel. *Front. Mater.* **2019**, *6*, 232. [CrossRef]
32. Yue, X.; Yang, Z.; Huang, L.; Zhang, L.; Li, J.; Xue, Z.; Pan, J. Passivation characteristics of ultra-thin 316L foil in NaCl solutions. *J. Mater. Sci. Technol.* **2022**, *127*, 192–205. [CrossRef]
33. Lv, J.; Luo, H.; Liang, T.; Guo, W. The effects of grain refinement and deformation on corrosion resistance of passive film formed on the surface of 304 stainless steels. *Mater. Res. Bull.* **2015**, *70*, 896–907.
34. Ebrahimzadeh, P.; Peral, L.B.; González-Martínez, R.; Mardaras, E.; Fernández-Pariente, I. Influence of Severe Surface Plastic Deformation Induced by Shot Peening on Microstructure and Corrosion Resistance of fine grained 316 L stainless steel. *Corros. Sci.* **2024**, *231*, 111988. [CrossRef]
35. LTaveira, V.; Montemor, M.F.; Da Cunha Belo, M.; Ferreira, M.G.; Dick, L.F.P. Influence of incorporated Mo and Nb on the Mott–Schottky behaviour of anodic films formed on AISI 304L. *Corros. Sci.* **2010**, *52*, 2813–2818. [CrossRef]
36. Dean, M.H.; Stimming, U. The electronic properties of disordered passive films. *Corros. Sci.* **1989**, *29*, 199–211. [CrossRef]
37. Zhao, M.; Wu, H.; Lu, J.; Sun, G.; Du, L. Effect of grain size on mechanical property and corrosion behavior of a metastable austenitic stainless steel. *Mater. Charact.* **2022**, *194*, 112360. [CrossRef]
38. Kale, A.B.; Kim, B.-K.; Kim, D.-I.; Castle, E.G.; Reece, M.; Choi, S.-H. An investigation of the corrosion behavior of 316L stainless steel fabricated by SLM and SPS techniques. *Mater. Charact.* **2020**, *163*, 110204. [CrossRef]
39. Jafarzadeh, S.; Chen, Z.; Zhao, J.; Bobaru, F. Pitting, lacy covers, and pit merger in stainless steel: 3D peridynamic models. *Corros. Sci.* **2019**, *150*, 17–31. [CrossRef]
40. Sun, D.; Jiang, Y.; Tang, Y.; Xiang, Q.; Zhong, C.; Liao, J.; Li, J. Pitting corrosion behavior of stainless steel in ultrasonic cell. *Electroch. Acta* **2009**, *54*, 1558–1563. [CrossRef]
41. Ernst, P.; Newman, R.C. Pit growth studies in stainless steel foils. I. Introduction and pit growth kinetics. *Corros. Sci.* **2002**, *44*, 927–941. [CrossRef]

metals

MDPI

Article

Processing, Microstructure, and Properties of Bimetallic Steel-Ni Alloy Powder HIP

Shenyan Huang [1,*], Chen Shen [1] and Victor Samarov [2]

1 GE Research, Niskayuna, NY 12309, USA
2 Synertech PM, Garden Grove, CA 92841, USA
* Correspondence: huangs@ge.com

Abstract: This work explores technical feasibility in hot isostatic pressing (HIP) manufacturing of an integral bimetallic component using steel and Ni alloy powder for supercritical carbon dioxide (sCO_2) turbomachinery. Lab-scale bimetallic HIP specimens using HAYNES® 282® and SS316L or SS415 powder are investigated in powder configuration, heat treatment, microstructure, and tensile properties up to 400 °C. Interdiffusion profiles at dissimilar alloy interfaces caused by HIP cycle is predicted by DICTRA simulations and validated by electron probe microanalysis (EPMA). The interdiffusion distance of most elements is around 100 μm, while C and N have a higher interdiffusion distance. Dense distribution of Ti-rich carbonitrides and alumina particles are found to decorate prior particle boundaries near joining interface on the 282 side, affecting tensile strength across interface as well as tensile failure location. A higher amount of excessive carbonitride formation near interface is observed in SS316L/282 than in SS415/282, which is consistent with the predicted greater degree of interdiffusion effect in SS316L/282. Typical HAYNES® 282® heat treatment condition is applicable to 282/SS316L and 282/SS415 combinations, resulting in a higher strength than cast CF8M and CA6NM. A pilot-scale bimetallic SS415/282 pipe is then demonstrated to show the promise of scaleup.

Keywords: hot isostatic pressing; powder metallurgy; steel; superalloy; dissimilar metal joining

1. Introduction

A high-temperature supercritical carbon dioxide (sCO_2) power cycle is being developed for concentrating solar-thermal power (CSP) application, providing more compact size and higher efficiency than a conventional steam power cycle [1]. A sCO_2 turbine that operates above 700 °C will require the use of γ' strengthened Ni-based superalloys for turbomachinery components at hot sections, leading to higher component cost than the lower temperature cycles that use structural steels. High cost is associated with manufacturing challenges and supply chain limitations of forging, casting, and machining large complex components in newer superalloys that are not broadly practiced such as HAYNES® 282® (Haynes International. Inc., Kokomo, IN, USA) and Inconel 740 H. Recently, alternative manufacturing modality of powder metallurgy (PM) near-net-shape (NNS) HIP has been demonstrated in a 10 MW sCO_2 turbine casing prototype, proving a significant reduction in component cost and levelized cost of electricity (LCOE) for CSP [2]. The present work further explores the technical feasibility in HIP manufacturing of an integral bimetallic component using steel and Ni alloy powder, which potentially further reduces powder cost by replacing Ni alloy partially with steel at lower temperature locations.

HIP diffusion bonding has been studied to join two similar or dissimilar metals in solid forms. Compared to conventional diffusion bonding that involves a uniaxial low pressure and limited plasticity, HIP diffusion bonding uses isostatic pressure in higher magnitude to increase diffusion rate and plasticity to achieve a sound metallurgical bonding that is free of voids [3]. The solid-state nature of the HIP process allows improved bonding strength compared to fusion welds (due to lack of heat affected zone and fusion zone) and joining of difficult-to-weld alloys (i.e., oxide dispersion strengthened alloys) [4–6]. In other papers, this process is also called HIP cladding where powder-to-solid or solid-to-solid HIP diffusion bonding forms a conformal layer that is thicker than what conventional coating processes could achieve [7–9]. The effect of pressure applied by HIP is reported to change precipitation kinetics, thus affecting the diffusion rate and grain structure at the interface of Fe-Al and CrMo steel [10]. HIP diffusion bonding also has fewer limitations in component geometry and size than conventional diffusion bonding. A variety of components in material combinations of Ni, Co, Ti, or steel alloys have been produced including valve body, well head, flywheel, extrusion dies, grinding roll for corrosion and wear-resistant applications, and turbine blisks [3,7–9,11–13]. Feasibility of HIP bonding Ni superalloy and Mo-Re refractory alloy, dissimilar Vanadium alloys, Cu alloy and Ni alloy, stainless steel and Ni alloy, and stainless steel and Ti alloy have been studied at lab-scale [14–17]. To the best of our knowledge, powder-to-powder HIP diffusion bonding has rarely been studied [18–20]. The powder-to-powder route was considered to be optimal due to the possibility of forming mechanical locks at the interface and local deformation of the particles at the interface for a better diffusion. Dissimilar powder bonding using spark plasma sintering (SPS) was studied for Inconel 718 and RR 1000 that resulted in no defects, intermetallic phases, and prior particle boundaries (PPB) at the bonding interface [21]. Despite the advantage of a shorter heating cycle in SPS to control kinetics of phase formation and grain coarsening, SPS does not have the scalability for large components with complex shapes, as the powder HIP approach has proven.

A HIP bonding strength with minimal penalties in mechanical properties highly relies on the alloy compositions, HIP cycle, post heat treatment, and interface preparation for cleanliness before HIP, which govern the resulting composition and phases, grain structure, and residual stress at bonding interface. Solid-to-solid HIP diffusion bonding requires low surface roughness, high cleanliness, and oxide-free bonding interface ensured by a physical or chemical interface preparation method prior to HIP [3,22]. The HIP temperature and post-HIP heat treatment shall be carefully balanced to reduce interface voids and prevent severe grain coarsening. Interlayer in solid state or melted state during the HIP cycle could be used at the joint interface to change the microstructure and composition, however, interlayer does not always benefit mechanical properties [23,24]. Similarly, powder HIP diffusion bonding requires a capsule containment of powder and sufficient outgassing followed by a leak check to ensure no leakage and contamination during HIP. Powder HIP has the advantage of finer grain size and thus potentially a wider window of HIP and heat treatment conditions. For dissimilar metal HIP, an interdiffusion zone up to hundreds of μm forms after HIP cycle, providing a transition in composition and coefficient of thermal expansion (CTE). This interdiffusion zone often prevents a sharp interface and reduces CTE mismatch-induced localized stress [11]. However, the formation of detrimental phases, dissolution or formation of carbides, and sharp concentration gradient within this interdiffusion zone may embrittle this region and significantly impact mechanical properties of the HIP bonding. Bonding interface between steel and Ni alloy could have complicated phases depending on the alloy combinations. For example, the interdiffusion zone between Ni-based Astroloy powder alloy and low carbon steel capsule was reported to contain five distinct sub layers of martensite, austenite, intense carbide precipitation zone, Fe-Ni rich superalloy, and Ni-Fe rich superalloy, whereas γ' precipitate size and fraction also varied within these layers [25]. Chemical potential and activity difference affects carbon diffusion at the interface, which in turn affects carbide type/size/fraction, phase of steel, and martensite phase transformation temperature. For example, dissimilar HIP bonding between tool steel (0.82% C) and martensitic steels (0.16–0.98% C) have distinct carbon profiles and diffusion distance at interface, where elevated carbon due to uphill diffusion occurs in one case [26]. PPB oxides and carbides formation due to carbon diffusion cause unsatisfactory properties in powder-to-solid HIP joints of SA508 and SS316L [27]. Additionally, phase transformation occurs during HIP or post heat treatments (such as austenite-martensite transformation in steel) which creates localized stress at the interface and increases the risk of cracking [8,26,28].

Thus, optimizing HIP diffusion bonding of dissimilar alloys requires a deep understanding of processing-structure-property relationships.

In the present work, bimetallic powder HIP between HAYNES® 282® and SS316L or SS415 is investigated computationally and experimentally. Using lab-scale HIP material, several configurations with different powder heat treatment conditions are studied. The interdiffusion profile at dissimilar alloy interface as a result of HIP cycle is understood by diffusion simulation using DICTRA and microstructure characterization. The strength of HIP bonding is measured by cross-interface tensile tests and compared against cast alloys. The role of carbonitride formed by carbon diffusion near interface is implied. A pilot-scale bimetallic pipe is then demonstrated to show some level of scalability.

2. Materials and Methods

Compositions of the studied alloy powders are listed in Table 1. Plasma atomized 282 alloy powder with size range of 20–180 μm and relative tap density of 67% was acquired from AP&C (Montreal, QC, Canada), a GE Additive company (Montreal, QC, Canada). Despite the higher cost than GA powder, PA powder has the advantages of improved cleanliness (i.e., N, O), highly spherical morphology, low trapped gas/porosity, and high yield in the powder size range that make it appropriate for PM HIP. Nitrogen gas atomized SS316L powder with size range of 44–106 μm and relative tap density of 60% was acquired from Carpenter Technology (Philadelphia, PA, USA). Two custom lots of argon gas atomized SS415 with a size range of 20–180 μm were acquired from Praxair Surface Technologies (Indianapolis, IN, USA). Relative tap density of the two lots were 64% and 60%, respectively. The first lot had lower carbon than expected, while the second lot had higher carbon, consistent with typical SS415.

Table 1. Actual powder alloy compositions (in wt.%).

Alloy	Ni	Co	Fe	Cr	Al	Ti	Mo	C	Mn	Si	N	O
SS316L powder	12.6	-	Bal.	17.82	-	-	2.32	0.021	0.96	0.71	0.09	0.015
SS415#1 powder	4.9	-	Bal.	12.9	-	-	0.74	0.002	0.61	0.08	0.003	0.063
SS415#2 powder	4.6	-	Bal.	12.2	-	-	0.76	0.028	0.75	0.03	0.01	0.031
282 powder	Bal.	9.95	0.95	19.5	1.44	2.09	8.26	0.067	0.15	0.084	0.006	0.007

To understand the interdiffusion and phase evolution at the interface between steel and 282 alloy during the HIP cycle, diffusion (DICTRA) simulations were employed to predict composition profiles across the interface and the resulting phases. In all simulations, the steel (either SS316L or SS415) was placed on the left and the 282 alloy was placed on the right, with a sharp interface in the middle initially. The length of either side was four mm, sufficient to cover the full diffusion distance of the (faster diffusion) interstitial elements, C and N, over the full HIP time-temperature cycle. A non-uniform grid was used to allow a fine resolution (grid spacing) within the interface region, at about one (center) to seven microns.

Austenite-martensite phase transformation temperatures upon heating and cooling in HIPed SS415 was measured by differential scanning calorimetry (DSC) using Netzsch STA 449C (Netzsch Instruments North America, Burlington, MA, USA). An yttria-coated, alumina-lined Pt-Rh crucible was used to hold a small specimen during DSC measurement. A heating and cooling rate of 10 °C/min and temperature up to 1500 °C were applied. Additionally, CTE and density of HIPed SS415 were measured by Netzsch DIL 402SU dilatometry up to 1200 °C, using a pin specimen with 25 mm length and 6 mm diameter. These results provide practical considerations on the post heat treatment condition and dimensional estimate after HIP.

Several conditions of steel/282 bimetallic powder HIP were created to study the interface microstructure and strength, as summarized in Table 2. Small HIP cans in SS304 with 19 mm diameter and 50 mm height were used for a microstructure study. Taller HIP cans in SS304 with 30 mm diameter and 180 mm height were used for mechanical property evaluation. One alloy powder was first filled in the bottom half of the can. Then the can containing half-filled powder was heat treated in a vacuum furnace at either 1000 °C or 1100 °C for 30 min, to allow slight sintering of powder at surface and forming an immobile powder surface. Then another alloy powder was filled in the top half, followed by standard powder vibrational packing. The powder vacuum heat treatment was shown to successfully maintain a sharp dissimilar powder interface during powder packing. The can was outgassed at both room temperature and an elevated temperature to achieve acceptable leak up rate before being pinched off. A helium leak check was conducted to ensure the integrity of the can prior to sealing. The same HIP cycle of 1204 °C/15 ksi/4 h was applied on all the studied conditions to fully densify powder, using either 6–45 H or 14–30 H HIP unit at American Isostatic Presses. During HIP cycle, temperature and pressure were ramped simultaneously to the set points with a 10 °C/min ramp rate. After the temperature and pressure soak, pressure was released, and a natural furnace cooling was performed. Examples of the small and taller cans after HIP are shown in Figure 1. The typical HAYNES® 282® heat treatment procedure was applied after HIP cycle, including solution treatment (1149 °C/1 h/WQ) followed the by two-step aging treatments (1010 °C/2 h/AC + 788 °C/8 h/AC).

(a)

(b)

Figure 1. Examples of a small can loaded with SS316L and 282 powder (**a**) and a taller can loaded with SS316L and 282 powder (**b**) after HIP cycle.

Representative microstructures in as-HIP and heat-treated states were obtained by high-resolution backscattered electron imaging using a Hitachi SU-70 FEG-SEM (Hitachi High-Tech America, Inc., Schaumburg, IL, USA) with a Bruker XFlash 6-60 SDD EDS detector (Bruker, Billerica, MA, USA). Composition profiles across interfaces were measured by Electron Probe Micro-analysis (EPMA) using the JEOL 8530F microprobe (JOEL USA Inc., Ashland, OR, USA). The analyses were made at 15 kV, 50 nA beam current, with a three micrometer beam diameter and three micrometer step increments.

Screw-threaded dog bone tensile specimens were machined from the base metal of SS316L, SS415, 282, and across steel/282 interface. The specimens had 3.5 mm gauge diameter, 19 mm gauge length, 6.35 mm grip diameter, and 50 mm total length. No issue was experienced in machining the steel/282 bimetallic specimens using the same machining procedure. Tensile tests were performed at room temperature, 200 °C, and 400 °C with 1–2 repetitions at the same test condition. A constant displacement rate of 0.5 mm/min was imposed for each test, and an extensometer was applied at the specimen gauge to measure strain evolution.

Table 2. The studied conditions of steel/282 bimetallic powder HIP.

Bimetallic Configuration	Powder Filled in Bottom Half Can	Powder Vacuum Heat Treatment	Powder Filled in Top Half Can	HIP Cycle	Solution and Aging Treatments	Tempering Treatment
282/SS316L	SS316L	1000 °C/30 min	282	1204 °C 15 ksi 4 h	1149 °C/1 h/WQ 1010 °C/2 h/AC 788 °C/8 h/AC	N/A
SS316L/282	282	1100 °C/30 min	SS316L			N/A
282/SS415#1	SS415#1	1000 °C/30 min	282			600 °C/1 h
SS415#1/282	282	1100 °C/30 min	SS415#1			600 °C/1 h
282/SS415#2	SS415#2	1000 °C/30 min	282			N/A

N/A is not applicable.

3. Results

3.1. Interdiffusion Simulation and Phase Prediction for HIP Cycle

The simulated composition profiles of SS316L/282 and SS415/282 are shown in Figures 2 and 3, respectively, for a simplified full HIP cycle of 4 h at 1204 °C, i.e., without the heating and cooling portions. The actual powder compositions in Table 1 (SS316L powder, SS415#1 powder, and 282 powder) were used. The alternate composition, SS415#2 powder, was also simulated but the results are not shown here due to their insignificant differences. After diffusion, steep slopes of some elements (Ti, Cr, C, N, Si) were developed as a result of phase formation at the joining interface. In particular, C and N peaks are associated with MC (as Ti (C,N)) phase formation. It is also noted that non-linear Si segregation profiles were predicted in SS415/282, even though the difference in the bulk compositions was small.

Equilibrium phases were simulated with the final compositions from the interdiffusion simulations for both SS316L/282 and SS415/282. To compare them against the microstructures from actual specimens, the phase calculations were made at the final aging temperature of 788 °C. It is clearly shown in Figure 4 that there was a greater influence of interdiffusion to SS316L/282, as shown by the γ' phase profiles at the 282 side from the interface. It is evident that a higher amount of carbide (MC) phase was predicted in the SS316L case than in the SS415 case, however, both show the excess carbide formation on the 282 side of the initial interface.

For a more quantitative comparison among elements and between the two materials (SS316L and SS415), interdiffusion distance ("IDZ width") was extracted from the simulated composition profiles. In Figure 5a, all elements show a characteristic trend of square root of time for diffusion-limited processes. As expected, the interstitial elements C and N have higher diffusion rates than the substitutional (metallic) elements. C and N diffusion is similar between SS316L and SS415. Among the metallic elements, Si is among the fastest elements. Ni (not shown in the plot) has an almost identical diffusion rate as Fe. Most of these elements have nearly the same diffusion rate between SS415 and SS316L, except for Ti and Cr, which are slower in SS415. Overall, the response of interdiffusion distance to HIP temperature is similar between SS316L/282 and SS415/282. The interdiffusion distance varies among elements, with most elements being at about 100 μm (between Ti and Fe), Si and Cr about twice wider, and C, N far exceeding all the formers.

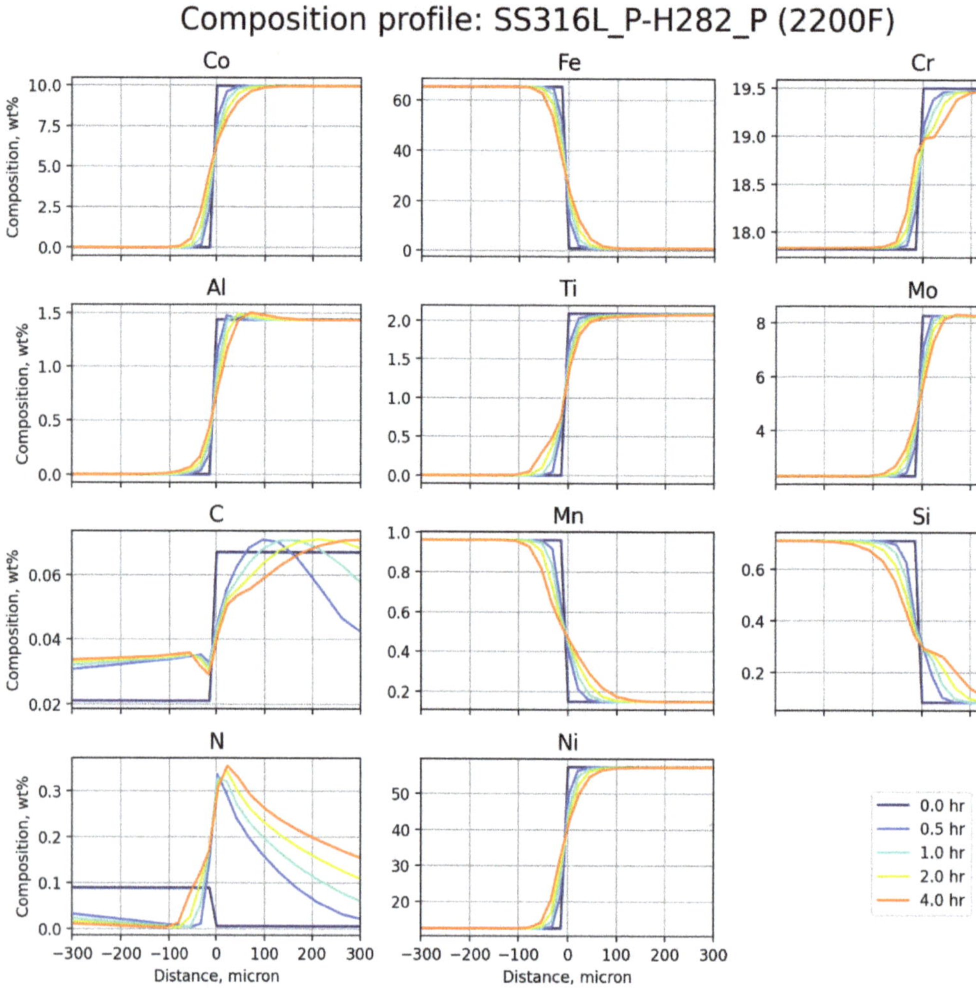

Figure 2. Simulated interdiffusion profiles between SS316L (**left**) and 282 (**right**) at 1204 °C (HIP temperature). Joining interface is in the middle.

3.2. SS415 Phase Transformation and Thermal Properties

Austenite/martensite phase transformation temperatures in SS415#1 were obtained by DSC analysis, as shown in Figure 6. Upon heating, the broad endotherm from 400 °C to 770 °C is likely due to overlap of the Curie magnetic transition (peak at 677 °C) and martensite to austenite transformation (peak at 707 °C). A smaller exotherm from 400 °C to 550 °C overlaps with the larger endotherm, probably due to an additional phase transition. Upon cooling, the exotherm with an onset of 398 °C and peak at 374 °C is likely due to austenite to martensite transition. Dilatometry analyses are presented in Figure 7. The ~0.2% linear shrinkage (~0.6% volumetric shrinkage) and density increase from about 657 °C to 718 °C is caused by martensite to austenite transformation, consistent with DSC observation. As a result, mean CTE and differential CTE both show a sharp drop around transformation temperature. Phase transformation temperature for SS415#2 was not measured, but martensite start temperature with the presence of higher carbon is anticipated to be slightly higher than SS415#1 (within 10 °C). The dimensional change due to phase transformation was considered in the HIP model to accurately predict the HIP shrinkage and reverse design the HIP tooling for a bimetallic pipe component (as described in Section 3.5).

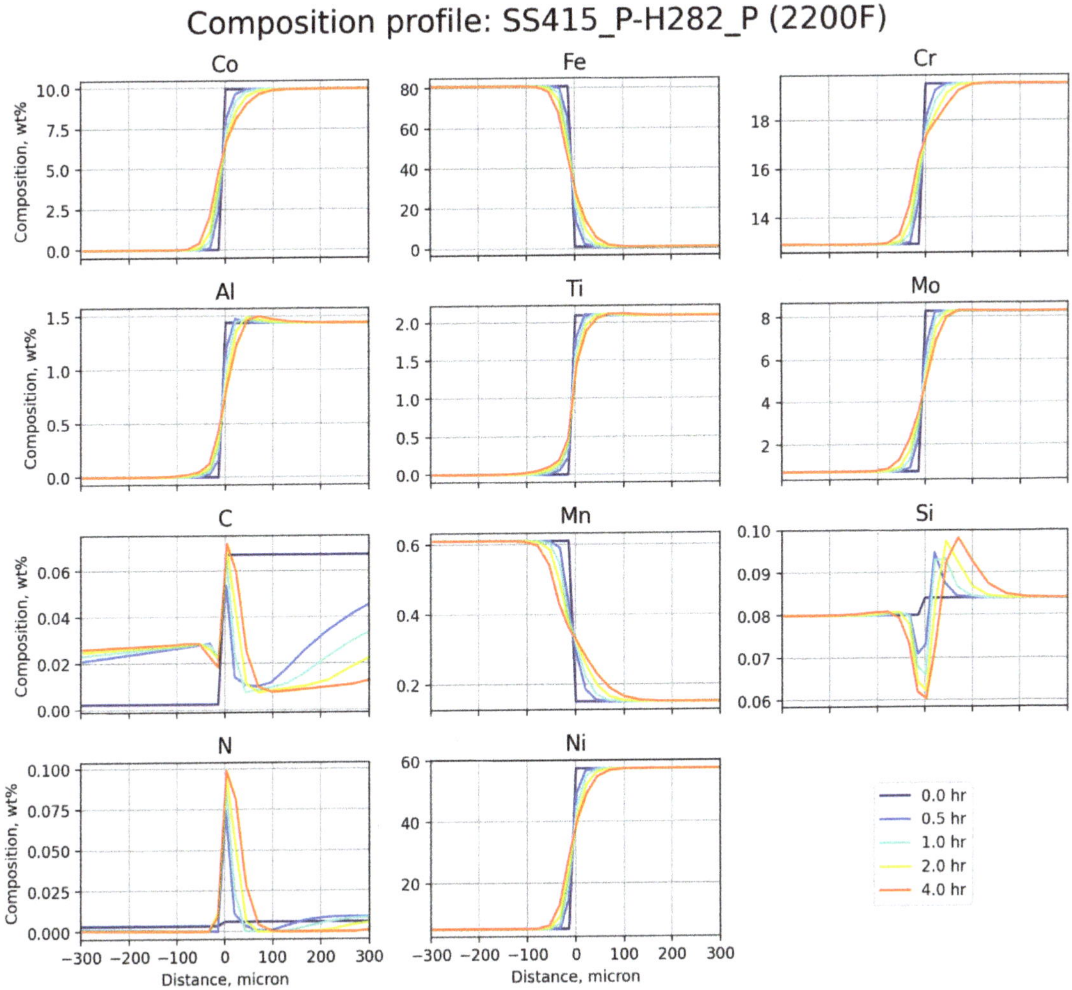

Figure 3. Simulated interdiffusion profiles between SS415 (#1 powder) and 282 (**right**) at 1204 °C (HIP temperature). Joining interface is in the middle.

3.3. Heat Treatment after HIP

The heat treatment of bimetallic powder HIP needs to satisfy the requirements of (1) being suitable for both 282 and SS316L or SS415 simultaneously, and (2) being suitable for the joining interface (transition region). The typical heat treatment conditions for the three alloys are listed in Table 3.

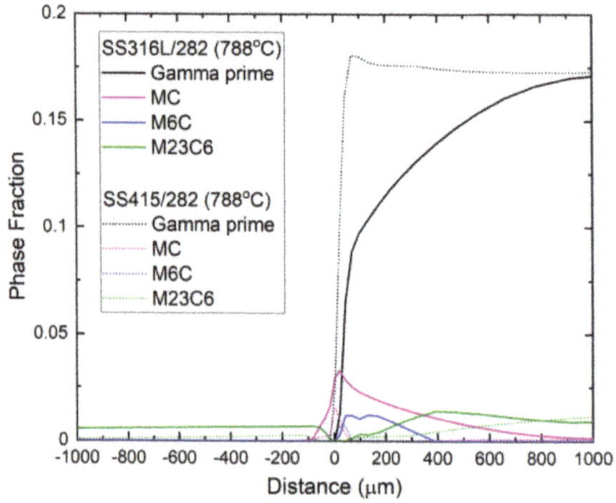

Figure 4. Simulated equilibrium phases at aging temperature 788 °C across interdiffusion zone of SS316L/282 and SS415/282.

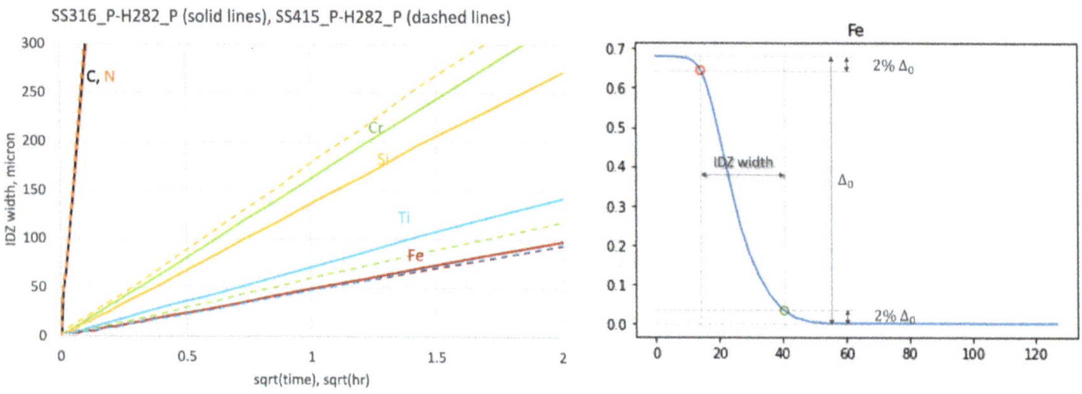

(**a**) (**b**)

Figure 5. (**a**) Comparison of the simulated interdiffusion zone (IDZ) width of selected elements for SS316L/282 and SS415/282 at 1204 °C (HIP temperature), (**b**) definition of IDZ width adopted in the plot.

Among these conditions, the solution/annealing of HAYNES® 282® (1121–1149 °C) and aging step 1 (1010 °C) are comparable to the annealing step of SS316L and the hardening step (i.e., austenitization followed by a quench to martensite) of SS415. Calculated phase diagrams (Figure 8) also suggest that 1010–1149 °C is well within the austenite temperature range of SS316L and SS415. Therefore, it is possible to replace the annealing step of SS316L and the hardening step of SS415 with the solution/annealing (and aging step 1) of HAYNES® 282®.

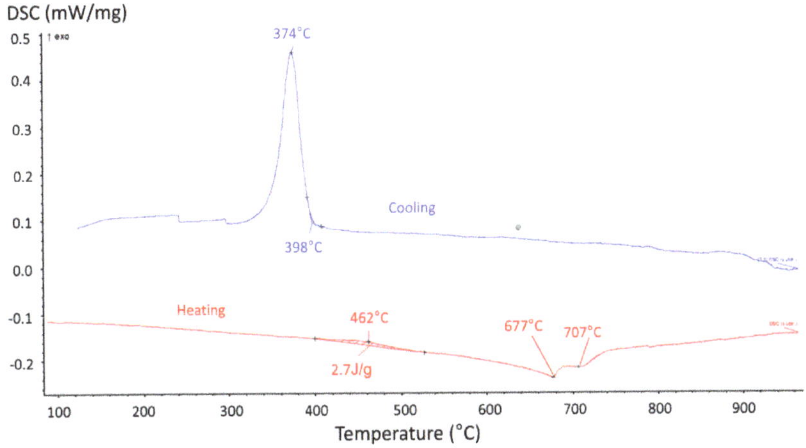

Figure 6. DSC heat flow vs. temperature for PM HIP SS415#1.

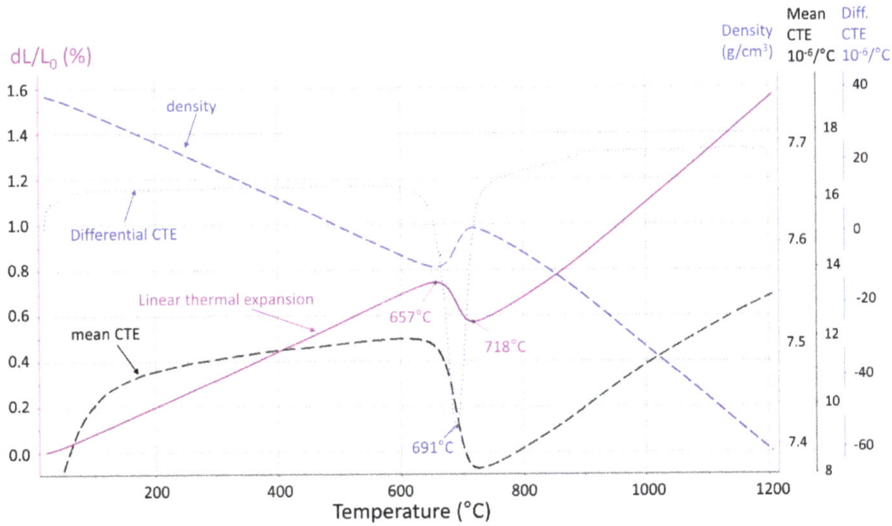

Figure 7. Thermal expansion, mean CTE, differential CTE, and density vs. temperature for PM HIP SS415#1 measured by dilatometry.

The main concern, however, is in regard to the 788 °C or 800 °C aging step of HAYNES® 282®. Firstly, SS316L is prone to formation of (Cr-rich) sigma and χ phases between 650–870 °C during an extended exposure, which can lead to reduction in mechanical properties and corrosion resistance (sensitization). The 788 °C/8 h or 800 °C/4 h aging step of 282 is within this window, as compared to a published TTT diagram for SS316L [29]. Several short-term heat treatment experiments were performed. Microstructural characterization verified that both aging conditions are safe to SS316L without causing brittle sigma and χ phases at grain boundaries. Therefore, solution and two-step aging of HAYNES® 282® was recommended for bimetallic HIP of SS316L/282 combinations.

Table 3. The typical heat treatment conditions for HAYNES® 282®, SS316L, and SS415 alloys.

HAYNES® 282®	SS316L	SS415
Solution/annealing: 1121–1149 °C, air cool (AC), Two-step Aging: 1010 °C for 2 h, AC; 788 °C for 8 h, AC. Or single-step Aging: 800 °C for 4 h, AC.	Annealing: 1040–1105 °C, AC.	Hardening: (except for +A) 950–1050 °C, AC or oil cool Tempering: 600–650 °C, AC or furnace cool ("soft annealed" +A); 650–700 °C, 600–620 °C, AC or furnace cool ("quench-tempered" QT650) 550–600 °C, AC or furnace cool (QT780) 520–580 °C, AC or furnace cool (QT900)

Regarding SS415, the 788 °C/8 h or 800 °C/4 h aging step is above the usual tempering temperature window of SS415 mentioned prior. According to calculations (Figure 8), the temperatures are already into the lower end of the austenitic region of SS415. Combined with DSC analysis on PM SS415 (Figure 6), the martensite to austenite transformation occurs at 707–724 °C. Thus, a viable way for SS415 appears to combine HAYNES® 282® heat treatment and a subsequent tempering step at a lower temperature 600 °C. A tempering time of 1, 2, and 3 h was tested. It was expected that this tempering step would not affect the microstructure of 282 since 600 °C is lower than the two-step aging temperature of 282 (788 °C). Microhardness was performed on small coupons in various conditions, including as-HIP state, solution and two-step aging with/without tempering, solution and 1-step aging with/without tempering. As-HIP, solutioned, and aged conditions all show high hardness values above 280 HV. With tempering, hardness drops to 230–235 HV. The tempered samples with different soak time all have similar hardness values. The two-step aging and one-step aging conditions did not show a big difference. Therefore, 600 °C/1 h tempering was downselected and applied to the 282/SS415#1 subscale HIP samples.

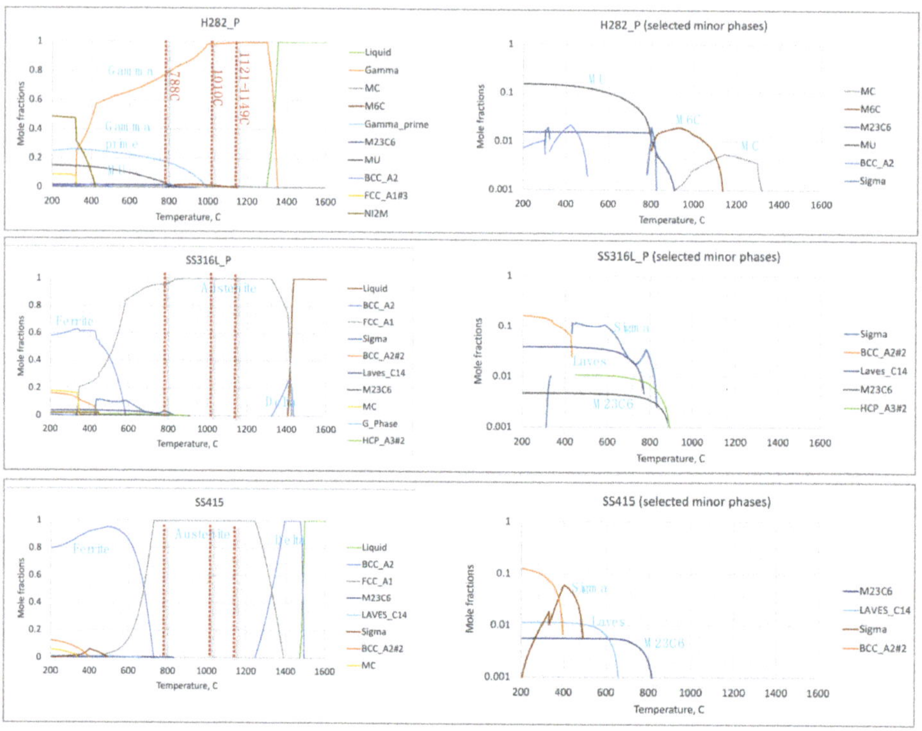

Figure 8. Calculated equilibrium phases.

3.4. Microstructure

Figure 9 displays representative microstructure in as-HIP state for 282/SS316L and SS316L/282 at the interface region. The applied powder vacuum heat treatment successfully created a slightly sintered "crust" of surface of the powder loaded in the bottom half of the HIP container. This crust effectively helped retain the flat shape of the powder surface during subsequent powder loading in the top half and vibrational packing. The well-defined interface observed in Figure 9 instead of mixed powder species is evidence of the sound process. The interface is relatively sharp and straight, as defined by the contours of individual powder. Consolidated 282 and SS316L on either side of the interface can be distinguished based on brighter and darker contrast in backscattered electron images. Close to the interface and on the 282 side, a dense distribution of dark-contrast discrete particles heavily decorate prior particle boundaries (PPBs) as well as inside powder particles. PPB particles in dark contrast also exist in the base metals, but in a much lower quantity and finer size. Comparing 282/SS316L and SS316L/282, powder loading sequence and vacuum heat treatment temperature that vary in the two configurations do not exhibit an obvious difference in the as-HIP microstructure. It can be seen in Figure 9 that very low density of dark contrast PPB particles are present in the base metal of 282 and SS316L, away from the ~100 μm interface region.

As-HIP interface microstructure of 282/SS415#1, SS415#1/282, and 282/SS415#2 is shown in Figure 10. A relatively sharp and straight interface was also visualized by the contrast difference of 282 and SS415. SS415 base shows a lath structured martensite within prior austenite grains. Compared to 282/SS316L alloy combination, a major difference in the interface microstructure of 282/SS415 alloy combination is a lower density of PPB particles. The location of these PPB particles is approximately 20 μm off the interface and within the 282 side. The first and second batch of SS415 powder differs significantly in carbon content, however, no noticeable difference was observed in the microstructure including interface PPB distribution, consistent with interdiffusion simulation results. It can be seen in Figure 10 that very low density of dark contrast PPB particles are present in the base metal of 282 and SS415, away from the ~100 μm interface region.

Figure 11 presents the interface microstructure of 282/SS415#1 in the fully heat-treated state. The secondary electron image (Figure 11a) is able to distinguish two types of fine particles in sub-micron size range that decorate PPB in a mixed mode: dark contrast particles in higher fraction and bright contrast particles in lower fraction. The EDS map (Figure 11b) further identifies the major chemistry of the two PPB particles: dark contrast being Ti rich carbonitrides and bright contrast being Al rich oxides. Similarly, these two PPB particles were also observed in 282/SS316L interface, despite a difference in quantity.

The microstructure after full heat treatment is similar to as-HIP state in terms of grain size, PPB particle size, distribution, and type at steel/282 interface. Applied heat treatment changes secondary phases in the 282 base metal by setting the right size for γ' precipitates and forming M23C6 carbides along grain boundaries, similar to secondary phase distribution in fully heat treated wrought HAYNES® 282®.

Composition profiles in as-HIP SS316L/282 and SS415/282 were measured with electron probe microanalysis (EPMA). Line profiles were made to traverse joining interfaces, with about 250 microns on either side (an example is shown in Figure 12). A beam diameter of three microns and step size of three microns were used. Each line profile produced about 167 quantitative spot measurements of eight elements: Fe, Ni, Co, Cr, Al, Ti, Mo, and Mn.

(a) (b) (c) (d)

Figure 9. Backscatter electron images of representative microstructure in as-HIP 282/SS316L (**a**,**b**) and SS316L/282 (**c**,**d**).

Figure 13 shows two SS316L/282 samples with a different can-filling procedure, described in Table 2. The difference in the measured composition profiles is not significant because of the same HIP condition, except that 282/SS316L shows a small step at the SS316L side (50–100 microns from interface) simultaneously in all elements (most noticeably Co, Fe, Ni). Repeat measurements at two additional locations in 282/SS316L confirmed that the variation was caused by local protrusion of 282 powders into the SS316L side during powder packing (i.e., local interface is not flat).

Diffusion simulations were performed with the actual HIP temperature-time condition. The simulated profiles are included in Figure 13 for each element. There is a good agreement with the EPMA data. In the regions far away from interface (about 200 microns, in both directions) the compositions are close to the 'bulk' composition of the powders. Since the simulations used actual powder compositions, the agreement with EPMA data in the bulk regions is not unexpected. Nevertheless, minor differences are still observed in Mn on the SS316L side and in Co and Al on the 282 side, while a greater difference in Cr is seen overall (and among the EPMA data sets). These differences are attributed to EPMA standard calibration and should not alter the conclusions here. A closer look at the Al profiles also indicates a small over-shoot near the interface (at the 282 side) in both simulation and EPMA.

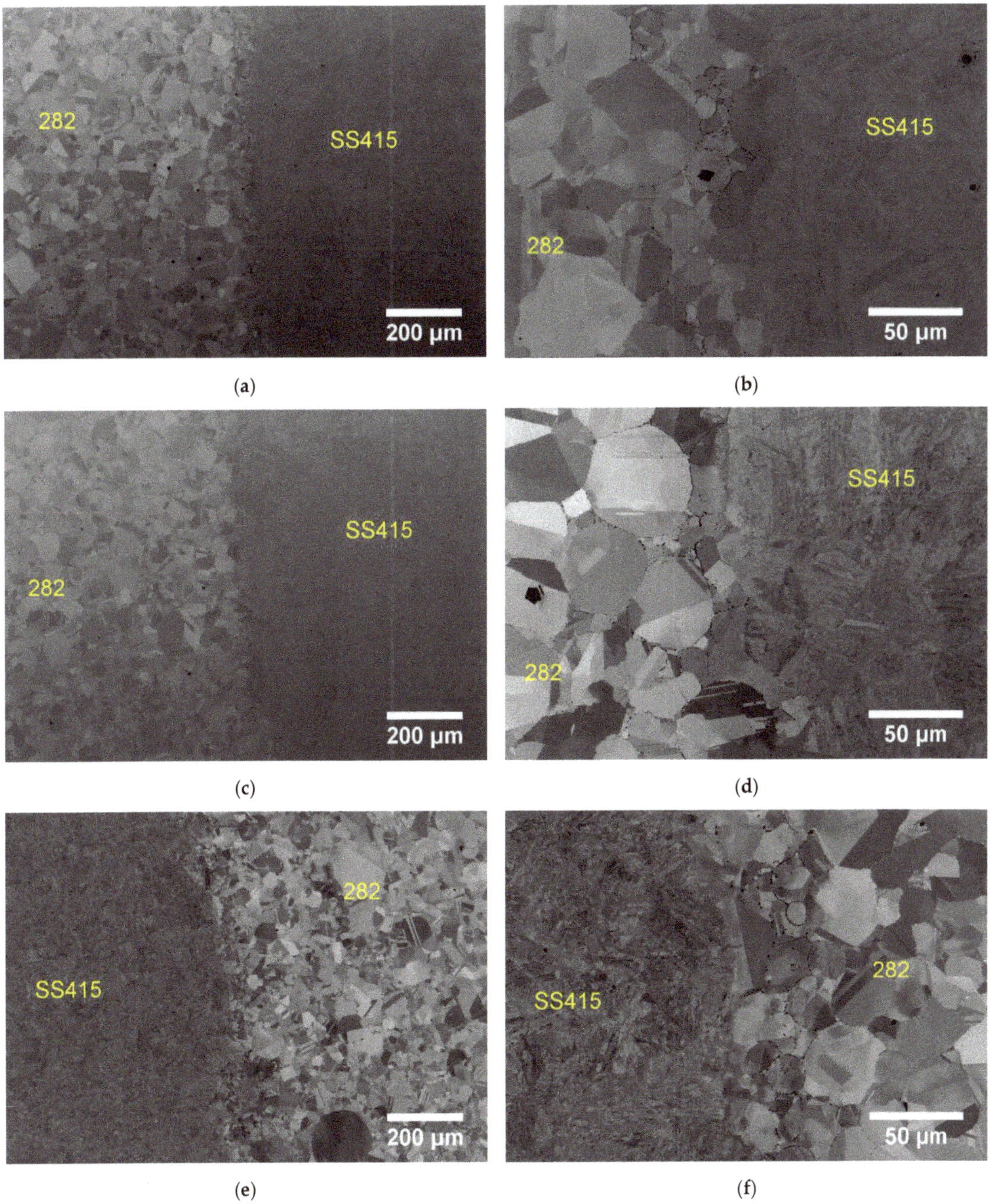

Figure 10. Backscatter electron images of representative microstructure in as-HIP 282/SS415#1 (**a**,**b**), SS415#1/282 (**c**,**d**), 282/SS415#2 (**e**,**f**).

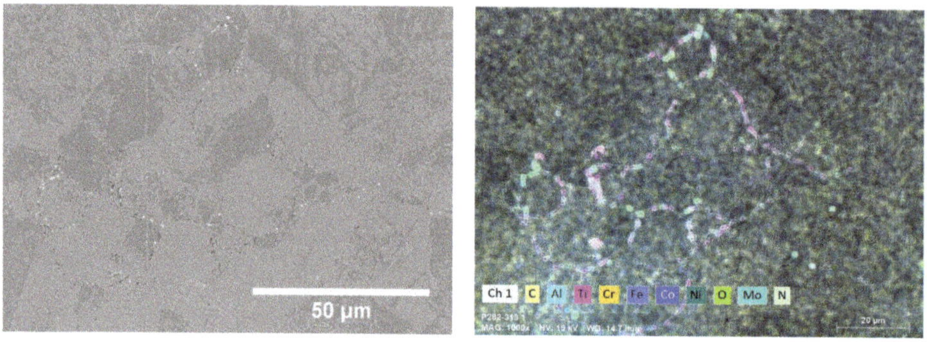

(**a**) (**b**)

Figure 11. Secondary electron image (**a**) and EDS map (**b**) of 282/SS415#1 interface region after full heat treatment.

Figure 12. Location of EPMA measurement on the sample SS316L/282 (red line). SS316L is on the left, 282 is on the right.

Although the interstitial elements C and N were not measured in EPMA and were unavailable for comparison with simulation, the EPMA's Ti data give some indirect evidence, as Ti is a forming element of M (C,N) carbide (carbonitride) phase. Near interface, the Ti data show a significant scatter because the EPMA's small incident beam (three micron) hits matrix phase or MC carbides, which have large difference in Ti compositions. The scatter is obvious because the local MC amount is high. The band of data scatter is approximately the width of the excess MC particles (thus C and N) formed at the interface. Figure 14 overlays these data with the simulated MC carbide fraction. The width of the interested region (~50 microns at interface) agrees very well between the EPMA and simulation. It is also noted that this is the latest evidence that the formation of these PPB particles is mainly on the 282 side.

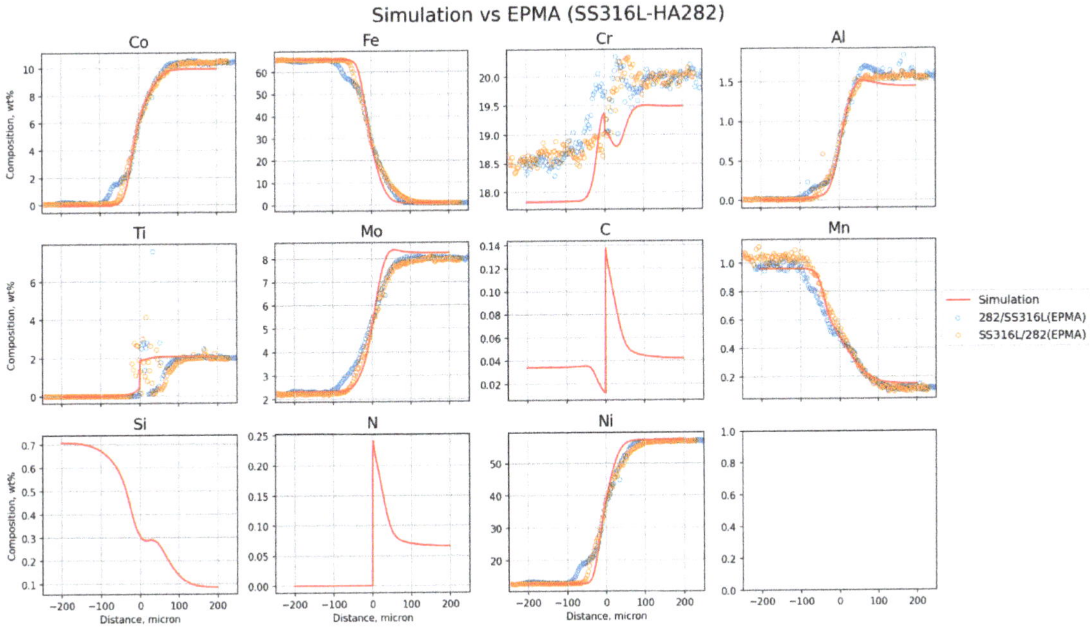

Figure 13. EPMA measurements and the simulated composition profiles between SS316L (**left**) and 282 (**right**) at 1204 °C (HIP temperature). Joining interface is at distance zero.

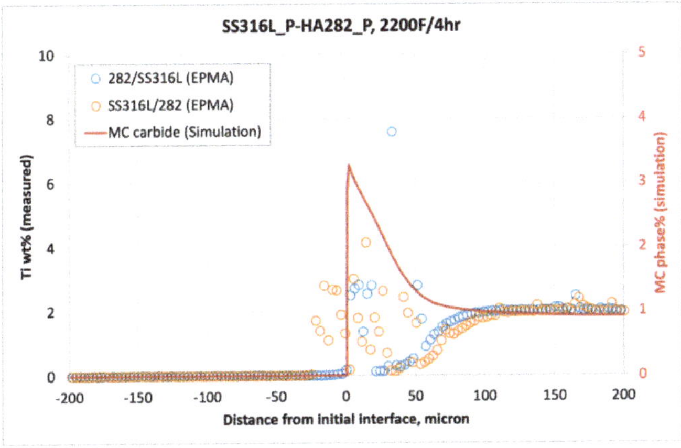

Figure 14. Overlay of scattered EPMA Ti data and simulated MC carbide profile. Data scatter is an indication of excess MC carbides near interface, agrees well with the simulation.

Figure 15 shows an overlay of the EPMA measurement and simulated composition profiles in SS415#1/282 samples. Similar to SS316L/282, the three SS415#1/282 samples using different can-filling procedures have only small difference in the measured composition profiles. A good agreement between EPMA data and simulation is also seen in SS415#1/282 samples. The amount of excess MC particles inferred from the scatter in EPMA Ti signal is much less in SS415#1/282 as compared to the SS316L/282 samples. This is consistent with the observation of less PPBs at SS415#1/282 joint interface from the SEM images reported previously.

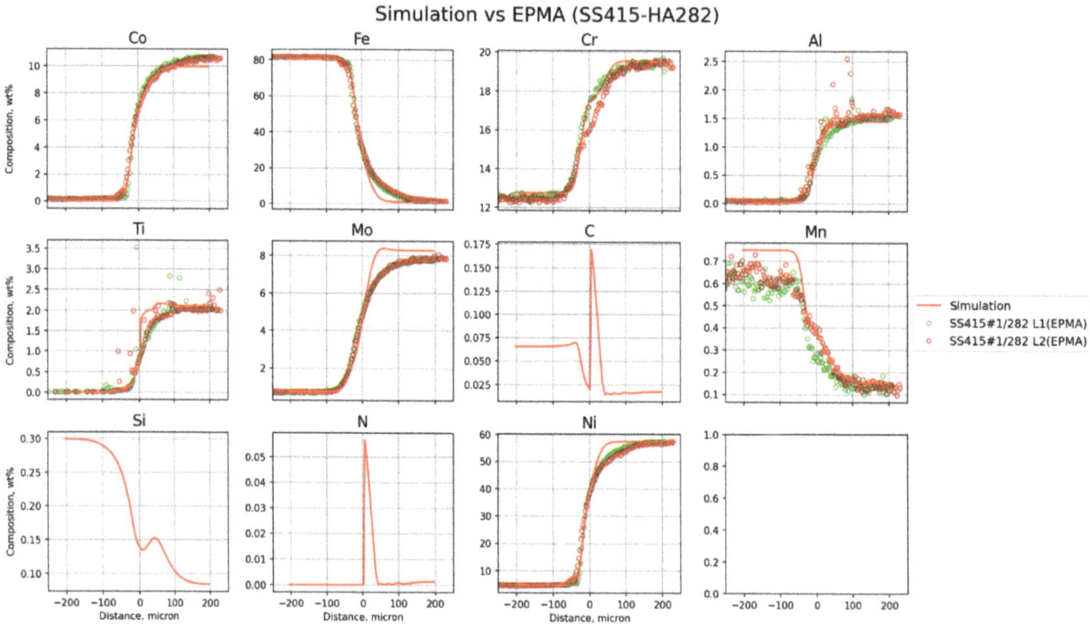

Figure 15. EPMA measurements (SS415#1/282, two line-profiles) and simulated composition profiles between SS415 (**left**) and Alloy 282 (**right**) at 1204 °C (HIP temperature). Joining interface is at distance zero.

3.5. Tensile Property

Figure 16 summarizes tensile properties of PM HIP 282 base metal, SS316L base metal, and two bimetallic cross interface configurations in comparison to wrought HAYNES® 282® and cast CF8M (cast SS316) [30]. Each data point represents one tensile test. Most 282/SS316L and SS316L/282 bimetallic specimens fail near or at the interface location (Figure 17a), except for two specimens tested at 400 °C. 282/SS316L and SS316L/282 bimetallic specimens show comparable yield strength as PM HIP SS316L base metal and cast CF8M. They have slightly lower ultimate tensile strength (UTS) than SS316L base metal, while still being comparable to cast CF8M. Consistent with our previous work [2], PM HIP 282 base metal shows slightly higher yield strength and UTS than wrought HAYNES® 282®, and its strength is much higher than SS316L base and bimetallic specimens. Powder loading sequence and vacuum heat treatment condition in the bimetallic HIP capsules do not generate any major differences in yield strength and UTS. Note that UTS of 282/SS316L and SS316L/282 bimetallic specimens are all significantly lower than yield strength of 282 base metal, indicating that the 282 half in the bimetallic specimens probably experienced little to no plastic deformation under the maximum tensile loading. Typical fracture surface of most bimetallic specimens that failed at interface reveals powder particle contours and fracture along PPBs (Figure 17b as one example). This implies that the dense distribution of Ti-rich carbonitrides along PPBs (Figure 9) as a result of interdiffusion during HIP cycle weakens the HIP bonding of dissimilar alloys and causes tensile failure.

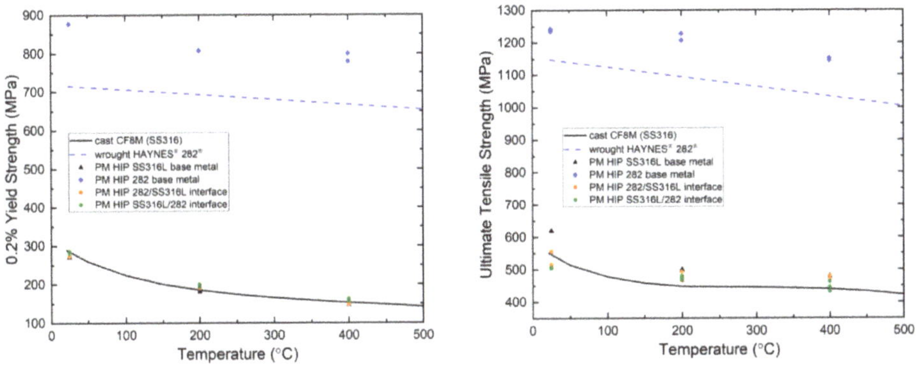

(a) (b)

Figure 16. 0.2% tensile yield strength (**a**) and ultimate tensile strength (**b**) of PM HIP SS316L, 282, and 282/SS316L interface in comparison with wrought HAYNES® 282® and cast CF8M (SS316).

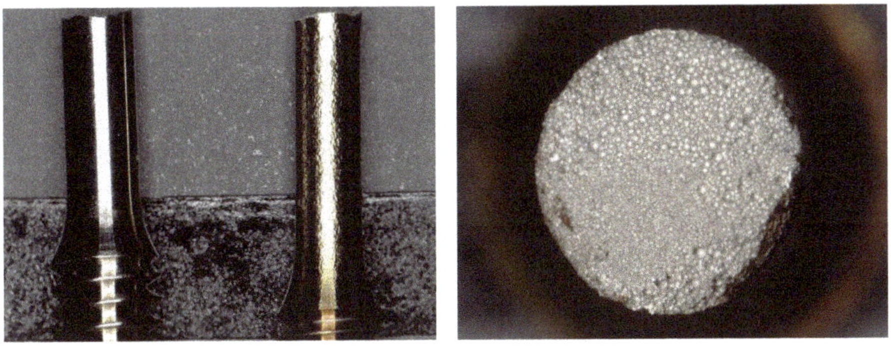

(a) (b)

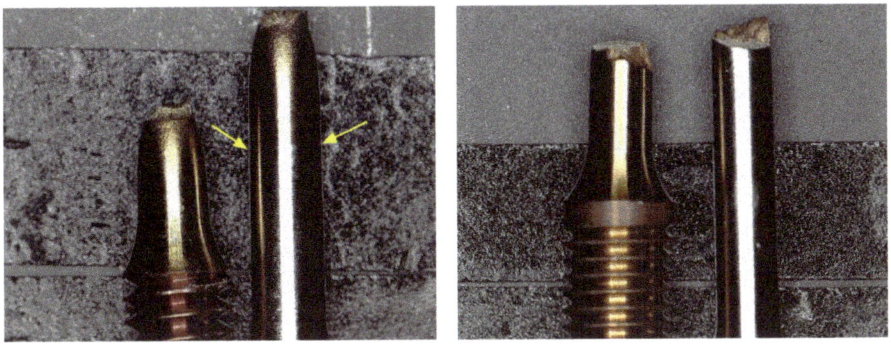

(c) (d)

Figure 17. Keyence optical images of tensile ruptured specimens and fracture surface: (**a**,**b**) SS316L/282 400 °C tested; (**c**) 282/SS415#1 400 °C tested (yellow arrows indicating the interface between two alloys); (**d**) SS415#2/282 400 °C tested.

Tensile results of 282/SS415 configurations are plotted in Figure 18. The first and second lot of SS415 powder with low and high carbon content result in quite different mechanical properties. Using the first powder lot, all the 282/SS415#1 and SS415#1/282 bimetallic specimens failed within the SS415 base metal, indicating that the interface strength was higher than SS415#1 base metal. Figure 17c shows one example of a 400 °C tensile ruptured specimen, where discoloration of SS415 due to oxidation and lack of discoloration in 282 clearly visualizes the interface location as pointed by yellow arrows. Additional metallography analysis on the longitudinal cross sections observed no microcrack or microvoid damage at the interface, confirming a good bonding. Yield strength and UTS of bimetallic specimens are slightly higher than SS415#1 base. It is believed that having a lower fraction of carbonitrides at 282/SS415 interface (Figure 10) than 282/SS316L interface (Figure 9) contributes to a higher metallurgical bonding that prevents interface rupture. Absent of interface failure, the difference in the powder heat treatments cannot be identified. The 600 °C/1 h tempering treatment added after the second aging step was intended to increase toughness, while strength of SS415#1 unfortunately dropped below that of cast CA6NM (cast SS415) [30] possibly without sufficient carbon content in the first powder lot.

The second SS415 powder lot results in substantially higher yield strength and UTS in SS415#2 base metal than the first powder lot, possibly due to higher carbon content. Using a standard 282 heat treatment without tempering, tensile strength of SS415#2 and SS415#2/282 both outperform that of cast CA6NM. However, the bimetallic specimens present a property penalty in yield strength and UTS compared to SS415#2 base metal, leading to tensile failures around interface. Interestingly, several test specimens had fractured surfaces in partial 282 and partial SS415#2 base metals (one example shown in Figure 17d). Not all the test specimens failed along the interface. The lower density of PPB precipitates near interface (Figure 10) probably reduces the chance of tensile rupture along these PPB networks.

Given the strength requirement of the target component, 282/SS415#2 alloy combination with typical HAYNES® 282® heat treatments are downselected for the dual alloy pipe demonstration. SS316L base metal and 282/SS316L interface do not have sufficient strength.

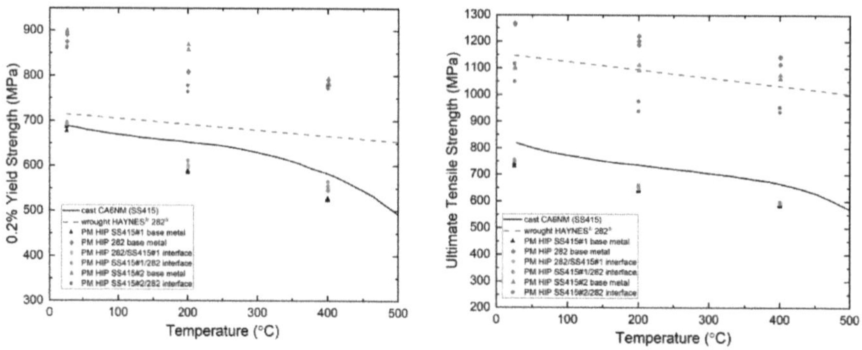

(a) (b)

Figure 18. 0.2% tensile yield strength (**a**) and ultimate tensile strength (**b**) of PM HIP SS415, 282, and 282/SS415 interface in comparison with wrought HAYNES® 282® and cast CA6NM (SS415).

3.6. Bimetallic PM HIP Pipe Demonstration

Leveraging lessons learned from the lab-scale study, a bimetallic SS415#2/282 pipe in 10.75 inch outer diameter, 1 inch wall thickness, 8.75 inch inner diameter, and 8 inch length was demonstrated to show the scale-up feasibility. The four technological challenges were to be met during this demonstration:

- Providing the dimensionally controlled location of the two alloys in the pipe;
- Formation of the firm and well-controlled interface between the two alloys;
- HIP both alloys to 100% density;
- Achieving the overall dimensions of the bimetal pipe.

To achieve these goals, the HIP capsule was designed by accounting for the different rheology of the two powders: 282 and SS415#2. A proprietary HIP model that considered non-uniform shrinkage during powder HIP was applied to back calculate the required HIP capsule dimension. In this case, a straight hollowed pipe was used as the HIP capsule for low cost and easy fabrication. To form a firm and well-controlled interface, the bottom half of the HIP capsule was first filled with 282 powder and then vacuum pre-sintered at 1204 °C to form a "crust" on the surface. Right before being loaded in the vacuum furnace, the HIP capsule was subjected to additional vibration to level the powder and the top was kept open for convenient inspection. After powder heat treatment, SS415#2 powder was loaded on top of 282 powder and the same HIP cycle (1204 °C/15 ksi/4 h) was applied. Figure 19a,b show the successfully HIPed capsule with a flat interface as a transition of OD and ID of the pipe. Figure 19c displays the measured dimension which nicely envelopes the target pipe dimension and allows for post machining. Powder packing density and material strength at HIP temperature determined more shrinkage in SS415#2 powder than 282 powder. The HIPed capsule was then machined into the final dimension of bimetallic pipe, which showed good machinability across the interface. Then, 282 heat treatment of solution and two-step aging was applied. Afterwards, the pipe was inspected by dye penetrant showing no cracks and defects particularly at SS415#2/282 interface (Figure 19d). SS415#2 material in the fill stems were examined by optical microscopy to confirm full consolidation through HIP and no thermally induced porosity (TIP). These results validate a successful demonstration of bimetallic HIP pipe that meets all the technical challenges.

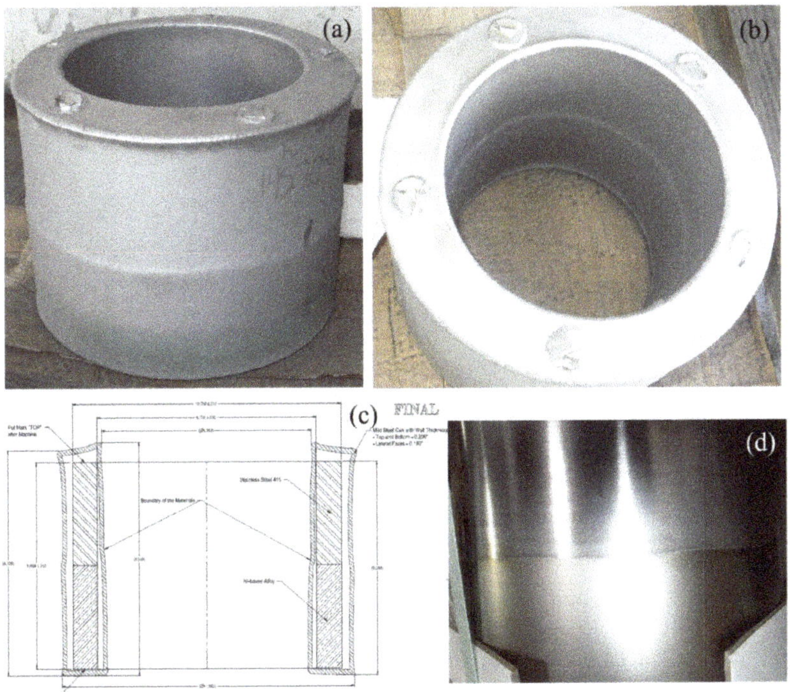

Figure 19. (**a**) Bimetallic pipe with HIP capsule material in as-HIP state: top half is consolidated SS415 powder, bottom half is consolidated 282 powder, interface is straight; (**b**) top-down view of inner diameter where interface is clearly seen; (**c**) measured post-HIP dimension (with capsule material) overlayed on top of target dimension of the bimetallic pipe; (**d**) closer view of interface after machining to the final dimension and applying heat treatment.

4. Conclusions

Feasibility in the HIP manufacturing of an integral bimetallic component using dissimilar alloy powder was investigated for 282/SS316L and 282/SS415 combinations. Interdiffusion profile at dissimilar alloy interface caused by the HIP cycle is predicted by DICTRA simulations and validated by EPMA analyses. Interdiffusion distance of most elements is around 100 μm, while C and N have a higher interdiffusion distance. Dense distribution of Ti-rich carbonitrides and alumina particles are found to decorate prior particle boundaries near the interface on the 282 side, affecting tensile strength across interface as well as tensile failure location. A higher amount of excessive carbonitride formation near interface is observed in SS316L/282 than SS415/282, consistent with the predicted greater influence of interdiffusion in SS316L/282. Typical HAYNES® 282® heat treatment condition is applicable to 282/SS316L and 282/SS415 combinations, resulting in higher strength than CF8M (cast SS316) and CA6NM (cast SS415). A pilot-scale bimetallic SS415/282 pipe has been then developed and demonstrated to show the promise of scale-up. The technological requirements of dimensional control of the two alloys and interface, defect-free interface by dye penetrant inspection, good machinability across interface, and microstructure are all met.

Author Contributions: Conceptualization, S.H.; methodology, S.H., V.S. and C.S.; validation, C.S. and V.S.; investigation, S.H., V.S. and C.S.; data curation, S.H.; writing—original draft preparation, S.H. and C.S.; writing—review and editing, V.S.; funding acquisition, S.H. All authors have read and agreed to the published version of the manuscript.

Funding: This work was funded by US Department of Energy, Energy Efficiency and Renewable Energy, Solar Energy Technology Office under award DE-EE0008996. GE Research provided cost share.

Data Availability Statement: The raw data supporting the conclusions of this article will be made available by the authors on request.

Acknowledgments: The authors thank Haynes International for HAYNES® 282® wire and AP&C (GE Additive) for plasma atomizing 282 powder. The authors would like to acknowledge technical support from Daniel Ruscitto, Marija Drobnjak, Robert Campbell, Chris Perlee, Chris Mclasky at GE Research.

Conflicts of Interest: Authors Shenyan Huang and Chen Shen were employed by the company GE Research. Author Victor Samarov was employed by the company Synertech PM (United States). All authors declare that the research was conducted in the absence of any commercial or financial relationships that could be construed as a potential conflict of interest.

References

1. Marion, J.; Macadam, S.; McClung, A.; Mortzheim, J. The STEP 10 MWe sCO2 Pilot Demonstration Status Update. In Proceedings of the ASME Turbo Expo 2022: Turbomachinery Technical Conference and Exposition, Rotterdam, The Netherlands, 13–17 June 2022; Volume 9: Supercritical CO_2; V009T28A034; ASME: New York, NY, USA, 2022. [CrossRef]
2. Huang, S.; Samarov, V.; Seliverstov, D.; Mortzheim, J.; Khomyakov, E. Near-Net-Shape HIP Manufacturing for sCO2 Turbomachinery Cost Reduction. In Proceeding of the 13th International Conference on Hot Isostatic Pressing: HIP'22, Columbus, OH, USA, 11–14 September 2022; Materials Research Proceedings 2023. Volume 38, pp. 11–16. [CrossRef]
3. Walker, R.M.; Roberts, D.J.; Rickinson, B.A. Component assembly using HIP diffusion bonding. In *Diffusion Bonding 2*; Stephenson, D.J., Ed.; Springer: Dordrecht, The Netherlands, 1991; pp. 261–269. [CrossRef]
4. Fu, H.Y.; Nagasaka, T.; Muroga, T.; Kimura, A.; Chen, J.M. Microstructural characterization of a diffusion-bonded joint for 9Cr-ODS and JLF-1 reduced activation ferritic/martensitic steels. *Fusion Eng. Des.* **2014**, *89*, 1658–1663. [CrossRef]
5. Hammelmann, K.H.; Buchkremer, H.P.; Stöver, D. HIP-diffusion bonding of ODS-materials by use of plasma sprayed encapsulation. In *Diffusion Bonding 2*; Stephenson, D.J., Ed.; Springer: Dordrecht, The Netherlands, 1991; pp. 250–260. [CrossRef]
6. Nagasaka, T.; Fu, H.; Muroga, T.; Tanaka, T.; Sagara, A.; Guan, W.; Nogami, S.; Hasegawa, A.; Serizawa, H.; Kimura, A.; et al. *Development of Dissimilar-Metals Joint of Oxide-Dispersion-Strengthened (ODS) and Non-ODS Reduced-Activation Ferritic Steels (IAEA-N-234)*; International Atomic Energy Agency (IAEA): Vienna, Austria, 2018.
7. Auger, J.P.; Raisson, G.; Pierronnet, M. Design and manufacture of engineered components clad by HIP. In *Hot Isostatic Pressing—Theory and Applications: Proceedings of the Third International Conference*; Koizumi, M., Ed.; Springer: Dordrecht, The Netherlands, 1992; pp. 229–234. [CrossRef]

8. Theisen, W. Hip cladding of tools. In *Proceeding of the 6th International Tooling Conference*; Bergström, J., Ed.; Karlstad University: Karlstad, Sweden, 2002; pp. 947–960.
9. Gauthier, M.; D'Amours, G.; Bernier, F. HIP Processing of Improved Tooling Materials for High-Productivity Hot Metal Forming Processes. In Proceeding of the 12th International Conference on Hot Isostatic Pressing: HIP'17, Sydney, Australia, 5–8 December 2017; Materials Research Proceedings 2019; Volume 10, pp. 85–91. [CrossRef]
10. Masahashi, N.; Hanada, S. Effect of pressure application by HIP on microstructure evolution during diffusion bonding. *Mater. Trans.* **2005**, *46*, 1651–1655. [CrossRef]
11. Butrim, V.N.; Beresnev, A.G.; Denisov, V.N.; Klyatskin, A.S.; Medvedev, D.A.; Makhina, D.N. Experience in HIP diffusion welding of dissimilar metals and alloys. In Proceedings of the 12th International Conference on Hot Isostatic Pressing: HIP'17, Sydney, Australia, 5–8 December 2017; Materials Research Proceedings 2019; Volume 10, pp. 65–72. [CrossRef]
12. Magerramova, L.A. Achievement of bimetallic blisks integrated dissimilar alloys for promising high temperature aviation gas turbine engines. In Proceedings of the 28th Congress of the International Council of the Aeronautical Sciences 2012, Brisbane, Australia, 23–28 September 2012; Volume 3, pp. 2457–2465.
13. Magerramova, L.; Kratt, E. Development of the Design and Technological Solutions for Manufacturing of Turbine Blisks by HIP Bonding of the PM Disks with the Shrouded Blades. In Proceedings of the 12th International Conference on Hot Isostatic Pressing: HIP'17, Sydney, Australia, 5–8 December 2017; Materials Research Proceedings 2019; Volume 10, pp. 39–46. [CrossRef]
14. Locci, I.E.; Nesbitt, J.A.; Ritzert, F.J.; Bowman, C.L. High temperature stability of dissimilar metal joints in fission surface power systems. *AIP Conf. Proc.* **2007**, *880*, 660–667. [CrossRef]
15. Zhang, L.W.; Fu, H.Y.; Chai, Z.J.; Zheng, P.F.; Wei, R.; Zhang, M.; Chen, J.M. Dissimilar-metal bonding of a carbide-dispersion strengthened vanadium alloy for V/Li fusion blanket application. *Tungsten* **2020**, *2*, 101–106. [CrossRef]
16. Xiao, Y.; Lang, L.; Xu, W.; Zhang, D. Diffusion bonding of copper alloy and nickel-based superalloy via hot isostatic pressing. *J. Mater. Res. Technol.* **2022**, *19*, 1789–1797. [CrossRef]
17. Dineshraj, S.; Manwatkar, S.K.; Gupta, R.K.; Bajargan, G. Development of hot isostatic pressing technology for joining multiple joints of dissimilar materials of stainless steels with IN 718 and Ti-6Al-4V. *Trans. Indian Inst. Met.* **2024**. [CrossRef]
18. Stavroulakis, E.; Bowden, D.; Irukuvarghula, S.; Garner, A.; Pickering, E.; Stewart, D.; Preuss, M. Characterisation of ferritic to austenitic steel functional grading via powder hot isostatic pressing. *Mater. Today Commun.* **2022**, *31*, 103442. [CrossRef]
19. Stavroulakis, E.; Irukuvarghula, S.; Pickering, E.; Stewart, D.; Preuss, M. Fundamental aspects of functional grading via powder hot isostatic pressing–Development of microstructure and diffusional processes. *Mater. Des.* **2022**, *215*, 110437. [CrossRef]
20. Zhu, W.; Liu, W.; Ma, Y.; Cai, Q.; Wang, J.; Duan, Y. Microstructural characteristics, mechanical properties and interfacial formation mechanism of tungsten alloy/steel composite structure fabricated by HIP co-sintering. *Mater. Des.* **2021**, *211*, 110127. [CrossRef]
21. Lister, S.; Blanch, O.L.; Fernandez, D.S.; Pope, J.; Baxter, G.J.; Bray, S.; Jackson, M. A Machinability Assessment of the Novel Application of Field-Assisted Sintering Technology to Diffusion Bond (FAST-DB) and Functionally Grade Dissimilar Nickel-Based Superalloys. *Metall. Mater. Trans. A* **2023**, *19*, 1–3. [CrossRef]
22. Larker, R.; Ockborn, J.; Selling, B. Diffusion bonding of CMSX-4 to UDIMET 720 using PVD-coated interfaces and HIP. *J. Eng. Gas Turbines Power.* **1999**, *121*, 489–493. [CrossRef]
23. Buchkremer, H.P.; Ennis, P.J.; Stover, D. Manufacture and stress rupture properties of hipped austenitic-ferritic transition joints. *J. Mater. Process. Technol.* **1999**, *92–93*, 368–370. [CrossRef]
24. Guelman, A.A. HIP diffusion bonding of intricate shape components made of light alloys and steels. In *Proceedings of the KWS Conference*; The Korean Welding and Joining Society: Daejeon, Republic of Korea, 2002; pp. 769–775.
25. Bassini, E.; Vola, V.; Lorusso, M.; Ghisleni, R.; Lombardi, M.; Biamino, S.; Ugues, D.; Vallillo, G.; Picqué, B. Net shape HIPping of Ni-superalloy: Study of the interface between the capsule and the alloy. *Mater. Sci. Eng. A* **2017**, *695*, 55–65. [CrossRef]
26. Frisk, K.; Luo, C.; Johansson, S.C.; Haglund, S.; Petterson, N.; Strandell, I. Compound materials by PM-HIP. *Powder Metall.* **2014**, *57*, 341–347. [CrossRef]
27. Le, J.; Yang, J.; Yin, H.; Samarov, V.; Gandy, D.; Lou, X. SA508 low alloy steel to 316L stainless steel dissimilar metal joint made by powder metallurgy hot isostatic pressing. *Mater. Sci. Eng. A* **2023**, *875*, 145060. [CrossRef]
28. Burlet, H.; Martinez, M.; Cailletaud, G. Microstructure and residual stresses issued from the bonding of an austenitic onto a ferritic steel by solid diffusion. In Proceedings of the 4th European Mechanics of Materials Conference on Processes, Microstructures and Mechanical Properties, Metz, France, 11 September 2001; Volume 11, pp. 157–164. [CrossRef]
29. Weiss, B.F.; Strickler, R. Phase Instabilities during High Temperature Exposure of 316 Austenitic Stainless Steel. *Metall. Trans.* **1972**, *3*, 851–866. [CrossRef]
30. Blair, M.; Stevens, T.L. *Steel Casting Handbook*, 6th ed.; Steel Founders' Society of America: Crystal Lake, IL, USA, 1995.

metals MDPI

Article

Deciphering Microstructures and Phases of Gas-Atomised Novel Al-Fe-Si-Cr-Ni Alloys

Bhaskaranand Bhatt *, Alessandra Martucci , Enrico Virgillito, Federico Gobber, Federica Bondioli, Diego Manfredi, Mariangela Lombardi and Paolo Fino *

Department of Applied Science and Technology, Politecnico di Torino, Corso Duca Degli Abruzzi 24, 10129 Turin, Italy; alessandra.martucci@polito.it (A.M.); enrico.virgillito@polito.it (E.V.); federico.gobber@polito.it (F.G.); federica.bondioli@polito.it (F.B.); diego.manfredi@polito.it (D.M.); mariangela.lombardi@polito.it (M.L.)

* Correspondence: bhaskaranand.bhatt@polito.it (B.B.); paolo.fino@polito.it (P.F.)

Abstract: Rapid solidification techniques, such as gas atomisation, have been widely implemented in metallic alloys/composites to increase solid solubility, avoid or mitigate segregation phenomena, and favour metastable phase formation to enhance performance. Particularly, gas atomisation can enhance the solid solubility of low diffusion coefficient elements like Fe, Ni, Mn, Zr, and Cr in the α-Al matrix, yielding metastable phases. As a result, Al alloys exhibit excellent strength at high temperatures. In this study, the AISI 304L alloy was employed to introduce Fe, Ni, and Cr elements into the AlSi10Mg alloy through gas atomisation, resulting in the formation of two distinct hypereutectic AlFe-based alloys: AlFe9Si8Cr2Ni and AlFe18Si8Cr5Ni2. Gas-atomised alloy powders were separated into different size fractions by sieving and characterised using X-ray diffraction, differential scanning calorimetry, optical microscopy, and scanning electron microscopy. Microstructural analyses revealed dendritic patterns with distinct phases, highlighting the influence of the alloying element content on the solidification processes. Furthermore, a synergic evaluation of the XRD and EDS analysis results allowed the identification of intermetallic phases and their distribution in the two systems.

Keywords: Al-Fe-Si alloys; iron correctors; sustainable alloy development; gas atomisation

1. Introduction

Since the 1900s, Al-Si alloys have stood out among the plethora of aluminium alloys thanks to their exceptional characteristics, including low density, good mechanical and physical properties, excellent castability, weldability, and high corrosion resistance [1]. Consequently, these alloys have found extensive application in the automotive and aerospace sectors, contributing significantly to critical components such as pistons, engine blocks, cylinder heads, and wheels [2]. Despite the undeniable advantages of Al-Si alloys, the growing demand for sustainable practices and the imperative to reduce environmental impact have highlighted a critical challenge in recycling Al-Si alloys: pervasive iron contaminations [3].

The presence of iron in Al-Si alloys increases properties at high temperatures and hardness but decreases corrosion resistance, fluidity, and plastic properties [4]. As previously reported, iron content exceeding 0.5% results in the formation of brittle intermetallic phases such as the Al_5FeSi (β) phase, which has an adverse impact on the mechanical properties of the alloy due to its needle-like morphology [5]. For instance, Lan et al. [6] reported that the Al_5FeSi (β) phase could be refined to the Al_8Fe_2Si (α) phase by increasing the cooling rate and Fe/Si ratio in Al-5Fe-xSi (X = 3.5–8 wt.%) alloys. This approach effectively mitigated the deleterious impact of the Al_5FeSi (β) phase and improved the alloy's thermal performance without sacrificing mechanical properties. Notwithstanding this, Al-Fe-Si phases are always discouraged, and finding a method to recycle Al-Si alloys with Fe contamination without decreasing the mechanical properties has become a growing concern. Since a reliable and economical method for removing iron from Al-Si alloys has not been known in metallurgical practice, the literature has turned to the exploration of iron correctors such as Mn, Ni, Co, Cr, Ti, V, and Mo [7].

Based on studies conducted in the literature, among the elements that can be used to correct the detrimental effects of iron in Al-Si alloys are Cr and Ni [7]. In particular, Cr demonstrated improved resistance at room and high temperatures without excessively burdening elongation at the break by precipitating intermetallic Cr phases acting as inoculant phases [8]. In addition, Cr is an inexpensive alloying element with low diffusion coefficients in aluminium that forms a solid solution with the α-Al matrix, exploiting solid-solution strengthening [9]. On the other hand, Ni in aluminium alloys is known to stabilise mechanical properties at elevated temperatures due to the low diffusivity in aluminium [10]. In addition, the presence of Ni in Al-Si alloys enhanced the corrosion resistance, decreasing the pitting rate under a long-term corrosion process [11]. Despite evidence of the beneficial aspects of Cr and Ni as iron correctors in Al-Si compositions, their content in the alloy is less than 1.0 wt.%. [7].

Obtaining these compositions with rapid solidification processes could increase the amount of alloying elements by extending their solid solubility limits in an Al-matrix due to the high cooling rates involved [12]. Among the techniques for producing metal powders that guarantee the highest cooling rates is inert gas atomisation (IGA) [13]. During IGA, a high-speed jet of inert gas is forced through tiny nozzles to disintegrate a molten stream of the alloy. The IGA process has become one of the most promising approaches for the development of metal compositions with a high content of alloying elements. In addition, the appeal of the IGA process could also be explained by its ability to produce a nearly spherical powder with a controlled particle size distribution suitable for the feedstock material of powder-based additive manufacturing (PB-AM) technologies [14].

To the best of the authors' knowledge, no available studies in the literature analyse the effect of high cooling rates in the IGA process on the amount of Fe, Cr, and Ni as alloying elements in Al-Si base alloys. To fill this literature gap, the present study focused on designing a novel Al-Fe-Si-Cr-Ni composition tailored for rapid solidification. In order to obtain the Al-Fe-Si-Cr-Ni compositions considering the pivotal aspect of sustainability, it was decided to start from ingots of two of the most widespread alloys for AM and conventional processes: AlSi10Mg (for Al and Si elements) and AISI 304L (for Fe, Ni, and Cr elements). In particular, these starting alloys were balanced to achieve Fe contents above and below its solubility limit in the aluminium matrix. Fe solid solubility in Al is 0.05 wt.% at thermodynamic equilibrium but can be increased up to 12.0 wt.% using rapid solidification processes [15]. For this reason, the compositions selected for the present study were AlFe9Si8Cr2Ni and AlFe18Si8Cr5Ni2. The two gas-atomised Al-Fe-Si-Cr-Ni powders were thoroughly investigated from a morphological and microstructural perspective, with attention paid to the role of the alloying element content and particle size.

2. Materials and Methods

2.1. Powder Production

The powders of the two Al-Fe-Si-Cr-Ni compositions were produced by our research group through a close inert gas atomisation process with PSI HERMIGA 100/10 VI lab scale atomiser equipped with a ceramic nozzle ø 2.5 mm. The AlFe9Si8Cr2Ni and AlFe18Si8Cr5Ni2 compositions were obtained starting from AISI 304L, and AlSi10Mg ingots in a ratio of 1:3 and 1:8, respectively. The chemical compositions of AlSi10Mg and AISI 304L ingots utilised as the initial feedstock for atomisation are reported in Tables 1 and 2.

Table 1. Composition of AISI 304L.

Elements	Cr	Ni	C	Mn	P	S	Si	N	Fe
Composition (wt.%)	18.0	8.0	0.02	1.50	0.03	0.03	0.41	0.08	Bal

Table 2. Composition of AlSi10Mg.

Elements	Si	Mg	Fe	Mn	Zn	Al
Composition (wt.%)	9.0–11.0	0.2–0.45	0.55	0.45	0.10	Bal

The ingots were subjected to vacuum induction melting in an Al_2O_3 crucible and were maintained at a superheat of 1000 °C for 1 h to ensure thorough and complete melting. In fact, as suggested by the ternary isothermal section of the Al-Si-Fe phase diagram, only the liquid phase is presented at 1000 °C [16]. After the melting process, the flow of molten metal into the atomisation chamber was initiated by opening the valve at the bottom of the crucible. As illustrated in Figure 1, the high-velocity jet of argon gas hits the molten metal as it falls, disintegrating it into numerous microdroplets that give rise to particles of different sizes. During gas atomisation, the disintegrated molten droplet is subjected to surface tension, drag, and buoyancy forces, and the balance of these forces results in an almost spherical morphology of the powder particles. The atomisation of the molten metal stream was carried out at a die pressure of 26 ± 1.25 bar, a gas flow rate of 2.55 ± 0.10 kg/ min, a gas-to-metal ratio of 3.19 ± 0.01, and a top pressure of 0.25 bar. In order to prevent surface oxidation of the particles, an inert high-purity argon atmosphere was used during the whole gas-atomisation process. In addition, due to the high reactivity of the Al-based compositions, a passivation step was carried out after gas atomisation to prevent explosivity issues.

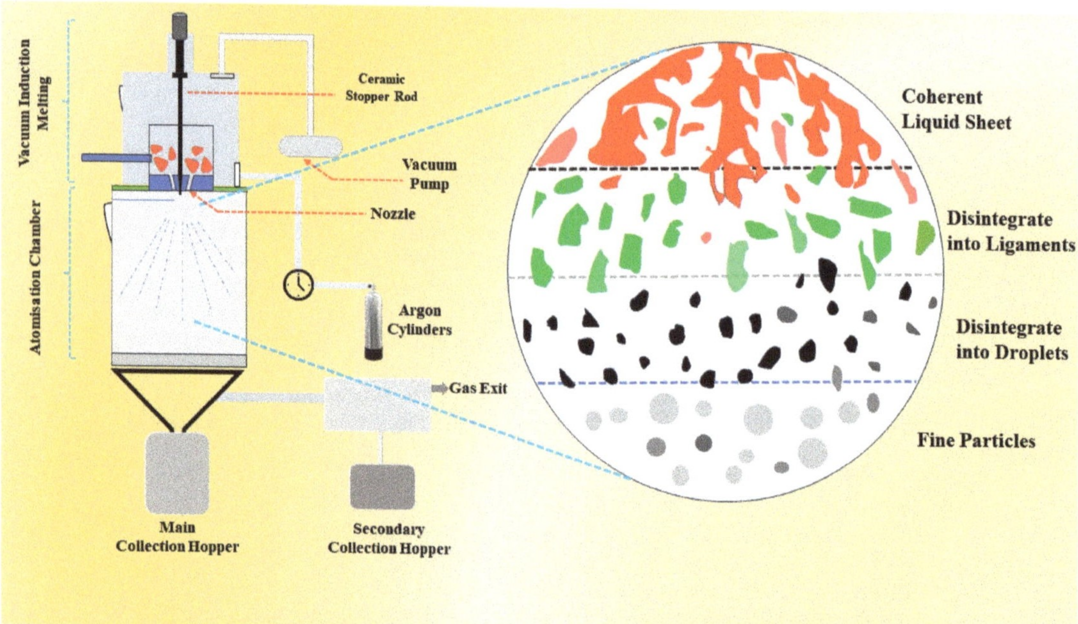

Figure 1. Schematic illustration of the Al-Fe-Si-Cr-Ni alloy development by close-coupled IGA.

2.2. Powder Characterisations

Before proceeding with the characterisation, the composition of the gas-atomised powders was assessed by performing the energy dispersive spectrometer (EDS) analysis on 10 particles for each system, as reported in Table 3.

Table 3. Chemical composition of the gas-atomised Al-Fe-Si-Cr-Ni powders.

Alloy	Fe	Si	Cr	Ni	Al
AlFe9Si8Cr2Ni (wt.%)	9.14 ± 0.30	8.60 ± 0.38	1.82 ± 0.10	0.77 ± 0.07	Bal
AlFe18Si8Cr5Ni2 (wt.%)	18.73 ± 0.45	7.42 ± 0.20	5.82 ± 0.28	2.12 ± 0.02	Bal

The particle size distribution (PSD) of the gas-atomised powders was assessed using the laser scattering technique with the Mastersizer 3000 analyser (Malvern Panalytical, Malvern, UK & Almelo, The Netherlands). To observe the particle morphology, the gas-atomised powders of the two compositions were analysed on stab by the Phenom XL scanning electron microscope (SEM, Phenom, Ambler, PA, USA). To verify the internal defects, the powders were mounted in hot resin and subjected to a dedicated metallographic preparation for a mirror-like finishing. Their cross-sections were observed by a Leica DMI500 optical microscope (OM, Leica Camera AG, Wetzlar, Germany).

The powders of both compositions were sieved following the ASTM B 215 standard using a Retsch AS 200 basic mechanical sieving shaker (Retsch, Haan, Germany), resulting in the following batches: <20 μm, 20–50 μm, and >50 μm. In order to investigate the effect of particle sizes and, thus, of different cooling rates on the morphology and composition of the present phases and the overall microstructure of AlFe9Si8Cr2Ni and AlFe18Si8Cr5Ni2 powders, a detailed microstructural analysis was conducted by using a Phenom XL SEM with a back-scattered electron detector (BSE) detector operating at a voltage of 15 kV. To investigate the different amounts of alloying elements on the microstructural properties of Al-Fe-Si-Cr-Ni systems, the 20–50 μm particle size range was selected. X-ray diffraction (XRD) analyses for phase identification and microchemical analyses by EDS detector for phase morphology and distribution investigation were conducted on the AlFe9Si8Cr2Ni and AlFe18Si8Cr5Ni2 20–50 μm batches. The XRD analyses were performed at 40 kV and 40 mA in a Bragg–Brentano configuration, using CuKα radiation. The XRD data was recorded in the 2θ range of 20 to 90° with a step size of 0.013° and a counting time of 45 s at each step. The phase identification procedure was performed by HighScore software (version 3.0d, Malvern Panalytical, Malvern, UK & Almelo, The Netherlands). EDS point, line, and map analyses coupled with SEM investigation in backscattered mode were employed to verify the presence of the phases identified from the XRD spectra and assess the differences in phase morphology and distribution as the alloying elements increase. The reaction sequences for AlFe9Si8Cr2Ni and AlFe18Si8Cr5Ni2 were determined through differential scanning calorimetry (DSC) analysis using the NETZSCH DSC 214 Polyma instrument (NETZSCH, Bayern, Germany). The DSC tests were conducted on the powder samples, each weighing approximately 75 mg. The heating rate was maintained at 10 °C/min, and the temperature ranged from 20 to 700 °C. At the same time, the cooling rate was maintained at 20 °C/min. The thermal analyses employed an Al_2O_3 crucible and a protective argon atmosphere.

3. Results

3.1. Particle Size Distribution

The PSDs of unsieved, gas-atomised powders are depicted in Figure 2a.

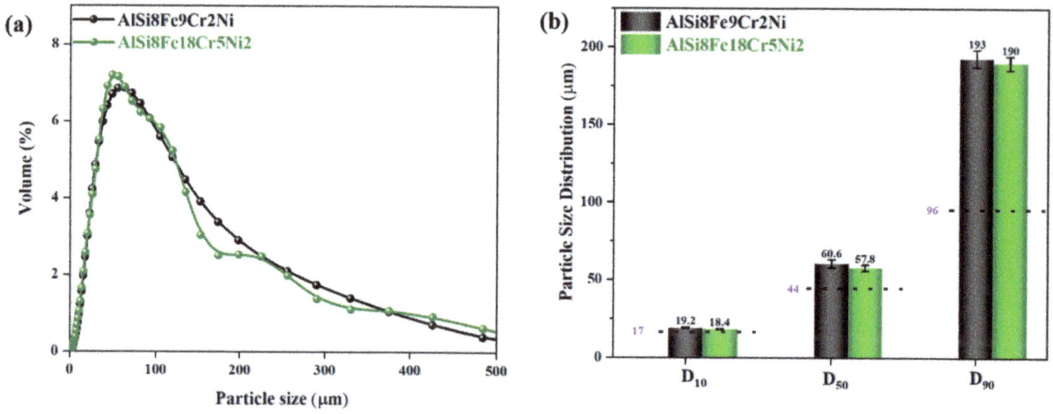

Figure 2. (**a**) Particle size distribution of gas-atomised powders. (**b**) D10, D50, and D90 percentiles of AlFe9Si8Cr2Ni and AlFe18Si8Cr5Ni2 compositions.

The particle size distribution curves appear asymmetrical for both compositions, with a sharp peak on the fine fractions and a marked tail towards the coarser fractions. The wide distribution can be attributed to the density and, thus, the viscosity in the liquid state of the two compositions. Indeed, the viscosity of the melt during gas atomisation influences the breakup of the molten metal stream into droplets, playing a crucial role in determining the size distribution of the metal powder particles obtained. Looking closer at the statistical parameters obtained from the cumulative PSDs reported in Figure 2b, it can be seen that the D10, D50, and D90 percentiles for AlFe9Si8Cr2Ni and AlFe18Si8Cr5Ni2 compositions are very similar to each other but appear much higher than the pure AlSi10Mg alloy (the AlSi10Mg references shown in dotted lines in Figure 2b pertain to a parallel study by the authors). In particular, D50 and D90 percentiles are more than 30 and 100%, respectively, higher than pure AlSi10Mg in both Al-Fe-Si-Cr-Ni alloys. This can be justified by the presence of Fe, Cr, and Ni as additional alloying elements that can alter the atomic arrangement and packing, impacting the viscosity. Based on studies conducted on the influence of Fe on the castability and properties of Al-Si alloys [3], it can be established that the addition of Fe as an alloying element in quasi-eutectic Al-Si alloys alters the local structure of the liquid, affecting the ease with which atoms move past each other. This leads to a drop in fluidity and, thus, an increase in melt viscosity. To the best of the authors' knowledge, the effects of Ni and Cr in Al-Si alloys have not been extensively investigated in the literature. Still, their limited presence in both compositions makes it plausible to assume that the presence of Cr and Ni may not significantly alter the increase in viscosity given by Fe. Higher viscosity leads to larger droplets during gas atomisation that tend to solidify more slowly, resulting in a larger final particle size [17].

3.2. Morphological Characterisation

During atomisation, the disintegrated molten droplets are subjected to surface tension, drag, and buoyancy forces, and the balance of these forces should result in an almost spherical morphology of the powder particles [18]. However, the use of the gas jet to break up the molten stream, turbulence that may occur close to the nozzle, or unintentional droplet collisions during gas atomisation may lead to powder defects that compromise the packaging, flowability, and spreadability of powder properties [19]. To verify the morphological conditions and the presence of internal defects in the AlFe9Si8Cr2Ni and AlFe18Si8Cr5Ni2 powders, optical analyses were conducted on the intact and cross-sectioned particles by SEM and OM (Figure 3).

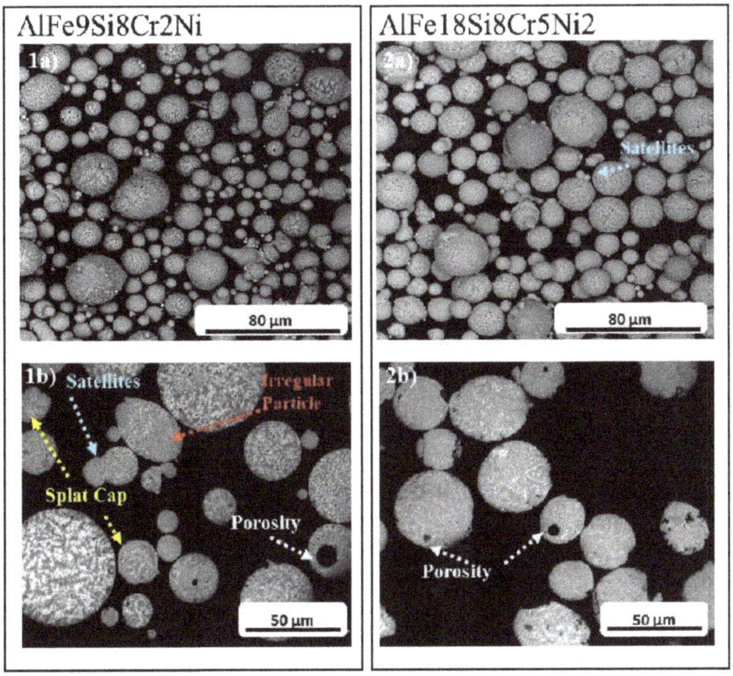

Figure 3. (**a**) SEM micrographs, and (**b**) optical images of cross-sectioned particles of gas-atomised AlFe9Si8Cr2Ni (**1**) and AlFe18Si8Cr5Ni2 (**2**) powders.

As it is possible to observe in the micrographs reported in Figure 3(1a,2a), the majority of gas-atomised particles exhibited a nearly spherical morphology in both Al-Fe-Si-Cr-Ni compositions. However, some particles displayed splat caps and irregular shapes. When a collision between a completely liquid droplet and a fully solidified particle occurs, splat caps like the ones indicated with yellow arrows in Figure 3(1b) could be formed [20]. In addition, unintentional collisions between droplets during solidification could result in shape deformation (indicated with red arrows in Figure 3(1b)) [20]. Such defects have detrimental effects on the powder properties; however, they are rarely present in the analysed powders. As shown in blue in Figure 3(1b,2a), the presence of satellites is also highly limited in both Al-Fe-Si-Cr-Ni compositions. Turbulence in the near-nozzle region that can occur during argon gas outflow at sonic velocity can generate the formation of tiny, already solidified particles, which can adhere to larger particles due to surface tension forces creating "outer satellites" or to molten droplets forming "embedded satellites" [21]. The poor occurrence of this defect in the analysed systems can be explained by the high viscosity of the molten stream due to the conspicuous presence of Fe in both alloys, which allows a slower, more stable gas atomisation process and the limited formation of ultra-fine particles. Additionally, there is no observable indication of open porosity in both compositions. Upon closer inspection of the particle surfaces of AlFe9Si8Cr2Ni and AlFe18Si8Cr5Ni2, a discernible dendritic pattern could be easily observed. Micrographs on the cross-sectioned powders reported in Figure 3(1b,2b) revealed limited levels of internal porosity. This defect is widespread in gas-atomised powders, as argon can become trapped in the molten droplet during chamber blowing [14]. The limited presence of internal porosity indicates the quality and stability of the atomisation process and makes the powders suitable for processing by PB-AM.

3.3. *Effect of Cooling Rate on Microstructural Evolution of AlFe9Si8Cr2Ni and AlFe18Si8Cr5Ni2 Powders*

The growth morphology of molten droplets during solidification is influenced by factors such as the solid–liquid interface's curvature, the solid–liquid interface's velocity, and the temperature gradient at this interface [22]. In order to investigate the effect of cooling rate and thus of particle size on the morphological evolution of the powders, AlFe9Si8Cr2Ni and AlFe18Si8Cr5Ni2 particles sieved below 20 μm, between 20 and 50 μm, and above 50 μm were observed in cross sections by SEM.

Both compositions show the presence of three different phases: one with a higher atomic weight (brighter in the BSE image), one with a lower atomic weight (grey in the BSE image), and one with the lowest atomic weight (darker grey/black in the BSE image). Starting from the analysis of the composition with the lowest content of alloying elements (AlFe9Si8Cr2Ni), it is possible to observe the outgrowth of fine dendrites from the liquid phase, a characteristic feature of rapid solidification [23]. Paying attention to the brighter phase, a significant microstructural variation can be observed as a function of the particle size. In particular, in the finer particles, cellular growth morphology can be observed (Figure 4(1a)), which changes into a dendritic one when the particle size increases (Figure 4(1b,1c)). The cooling rate of a molten droplet is intimately related to the heat extraction rate from the liquid droplet by the atomising gas [24]. Heat extraction from a droplet is the culmination of conductive heat transfer from the core of the droplet to the surface, convective heat transfer by the atomising gas from the droplet's surface, and heat transfer by radiation from the solidifying droplet. Thus, the heat transfer rate during solidification of molten droplets depends on the droplet size. Rapid heat transfer occurs in fine droplets, whereas heat transfer is comparatively sluggish for coarser droplets. Therefore, the droplet size is directly related to its cooling rate, which, in turn, influences the solidification microstructure [25]. Generally, there is a limited solute redistribution in front of the solid–liquid interface under high cooling rates. As a result, the finest particles are characterised by a globular microstructure that has not had sufficient time to evolve and create the dendritic morphology. This could explain the reason why, with increasing size, the dendritic structure appears progressively more evident. It can also be noted how the darker (black-like) phase seems not to be present in the finer particles and increases in quantity as the size increases. To name these phases, please refer to the EDS analysis results in Section 3.4.2.

Observing the BSE micrographs of AlFe18Si8Cr5Ni2 particles, it is evident that the solidification front, specifically dendritic solidification, remains consistent regardless of the particle size. This lower susceptibility of the morphology to particle size can be traced back to the conspicuous presence of iron in the alloy, which leads to an increase in the viscosity of the molten stream and, thus, to higher stability of the gas atomisation process. However, with increasing particle size, it is possible to note that the discontinuous dendrites become increasingly continuous, and their dimensions increase. In particular, the mean length of primary dendrites exhibited fluctuation, ranging from 4 to 7 μm for particles with diameters below 20 μm, 10 to 15 μm for particles with diameters 20–50 μm, and 15 to 20 μm for particles with diameters above 50 μm. This can be attributed to the lower cooling rate and, thus, the longer solidification time that characterises coarser particles and allows dendrites to grow. Furthermore, some particles exhibit multiple nucleation sites within a single particle (indicated in red in Figure 4(2a)). These nucleation sites are located on the powder's outer surface, which may be related to the strong thermal convection prevailing in that area.

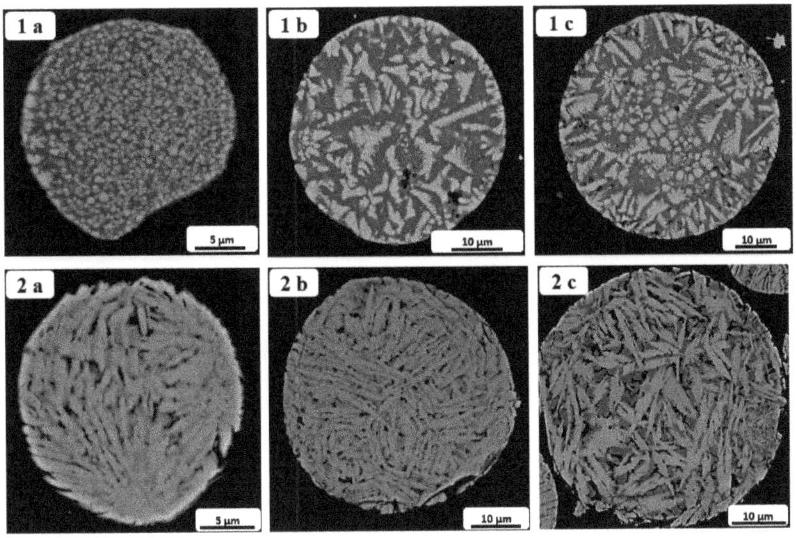

Figure 4. SEM micrographs of cross-sectioned particles of gas-atomised AlFe9Si8Cr2Ni (**1**) and AlFe18Si8Cr5Ni2 (**2**) powders ((**a**) <20 µm (**b**) 20–50 µm (**c**) >50 µm).

3.4. Compositional Characterisation

In order to investigate the amount and the distribution of the alloying elements for the two compositions, XRD analyses for phase identification and microchemical analyses by EDS for phase distribution investigation were conducted on AlFe9Si8Cr2Ni and AlFe18Si8Cr5Ni2 powders. Based on the above-described microstructural variations upon different particle sizes of the AlFe9Si8Cr2Ni composition, it was decided to use the median range of 20–50 µm as a representative sample for the compositional analysis of the powders.

3.4.1. Phase Identification through XRD

XRD analysis was used to determine how different amounts of alloying elements in the AlFe9Si8Cr2Ni and AlFe18Si8Cr5Ni2 systems affect the presence and amount of intermetallic phases. Analysing the XRD spectrum of AlFe9Si8Cr2Ni shown in black in Figure 5, it was possible to identify the most intense peaks related to the fcc-Al constituting the matrix of the system, other peaks mainly attributable to two intermetallic phases, $Al_{95}Fe_4Cr$ and $Al_{13}Fe_4$, and two weak peaks associable with the presence of Si diamond cubic. Some smooth peaks can be detected in the 29–35° range, but due to their low intensity, the software was not able to provide a reliable attribution. However, the literature pertaining to Al-Fe-Si alloys reported XRD peaks in the 20–35 range for α-Al_8Fe_2Si and/or β-Al_5FeSi phases [6,26]. In particular, Lan et al. [6] reported that the eutectic α-Al_8Fe_2Si phase is favoured in Al-Fe-Si compositions as the cooling rate and Fe-Si ratio increase. Based on the rapid cooling rates involved in the gas atomisation process and the high Fe-Si ratio of the AlFe9Si8Cr2Ni system (around 1.25), it is reasonable to assume that the eutectic phase may occur. However, further analyses are required to verify this assumption. According to the XRD spectrum analysis, it can be stated that only Si and Al are present as pure elements. At the same time, Fe reacts in its entirety with Al and other alloying elements, and Cr reacts preferentially with Al and Fe. Peaks related to Ni or Ni-containing phases were not identified. Still, it is not possible to definitively state if the absence can be attributed to the presence of Ni completely in solid solution or if it is related to the low amount of the element in the composition as quantities below 1.0 wt.% are not detectable with XRD analysis.

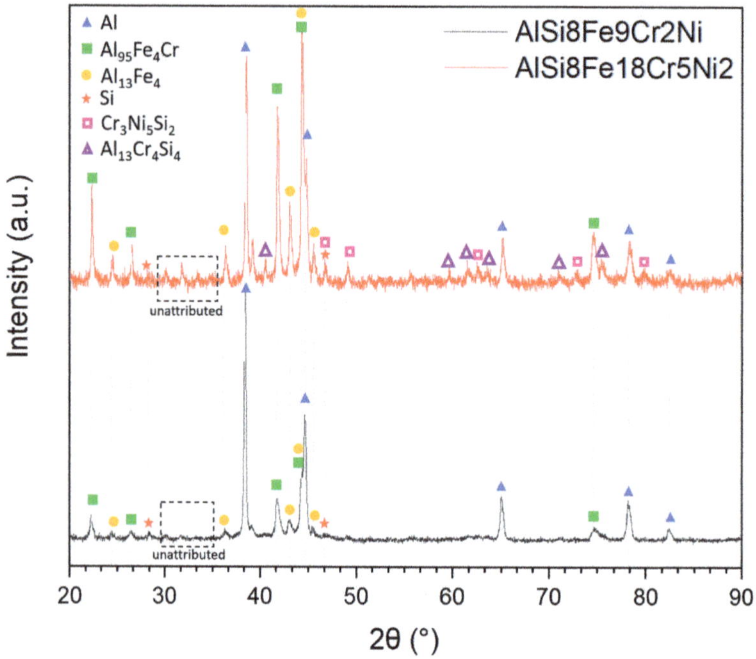

Figure 5. XRD patterns of 20–50 μm sieved AlFe9Si8Cr2Ni and AlFe18Si8Cr5Ni2 powders.

Moving to the observation of the XRD spectrum of AlFe18Si8Cr5Ni2 displayed in red in Figure 5, it can be noted that the phases detected in the system with a lower amount of alloying elements are still present, but the peak relative intensity is different, and the spectrum overall appears much noisier and with a massive presence of additional peaks. Looking at the phases already identified in the AlFe9Si8Cr2Ni system, it is evident that the presence of $Al_{95}Fe_4Cr$ and $Al_{13}Fe_4$ is more conspicuous. In fact, the dominant peak in the AlFe18Si8Cr5Ni2 XRD spectrum is no longer the Al (111) peak (38.37°) but is the one at 44.18° given by the co-presence of the main peak of the $Al_{95}Fe_4Cr$ phase and the third most intense peak of the $Al_{13}Fe_4$ phase. Furthermore, it can be seen that Si is no longer present only as a pure element but seems to react with Al, Cr, and Ni, resulting in two other intermetallic phases not identified in the AlFe9Si8Cr2Ni system: $Cr_3Ni_5Si_2$ and $Al_{13}Cr_4Si_4$. The appearance of the peaks relative to the $Cr_3Ni_5Si_2$ phase is in line with the higher amount of Ni in the composition that makes it detectable by XRD analysis and suggests that Ni reacts preferentially with Cr and Si rather than with the Al matrix or the second most abundant element in the alloy (Fe). In addition, the smooth peaks detected in the 29–35° range of the AlFe9Si8Cr2Ni system appear more intense in the AlFe18Si8Cr5Ni2 spectrum. Still, due to the high signal noise and the co-presence of numerous peaks referable to Al,Fe and Al,Fe,Si-containing phases in that angle range, it was not possible to provide a reliable attribution. A more accurate attribution of unassigned peaks has been postponed until microstructural analyses have been conducted.

3.4.2. Compositional Analysis by EDS

Scanning electron micrographs obtained with the BSE detector revealed, for both compositions, dendritic structures with higher atomic weights and an interdendritic region with a higher atomic weight. Looking closer at AlFe9Si8Cr2Ni particles, the interdendritic region appears to consist of an extended heavier phase (grey) and a brighter, less diffuse phase (black). To name the phases identified in the BSE micrographs, an EDS compositional analysis was conducted.

From the EDS maps reported in Figure 6, it is possible to note that the dendritic structures are clearly enriched in Fe. In addition, the distribution of Fe matches the Cr-enriched area. This result appears to be in line with the $Al_{95}Fe_4Cr$ phase found by the XRD analysis. Based on the low amount of Ni, its distribution is not reliably recognised by the EDS maps. Furthermore, Si and Al appeared widely present over the entire analysed surface, and for this reason, it was decided not to report their maps. To investigate the composition of the grey interdendritic areas, a line EDS analysis was performed. The latter recognised an alternation of Si and Al resembling the Al-Si lamellar structure typical of AlSi10Mg. These Al-Si alternations may support the assumption of the presence of Al-Fe-Si phases (also eutectic) made on unassigned peaks identified in the XRD pattern. In addition, their sub-micrometric dimensions could explain the low intensity of the relative XRD peaks.

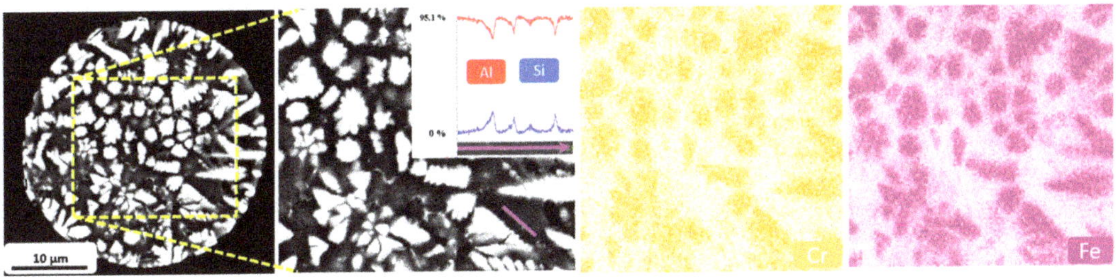

Figure 6. Cross-sectional SEM-EDS maps on cross-sectioned AlFe9Si8Cr2Ni powder with an SEM-EDS line scan in the interdendritic region (grey zone).

Observing the EDS maps reported in Figure 7, it is possible to note that in line with the results obtained for the Al-Fe-Si-Cr-Ni with a lower amount of alloying elements, also in AlFe18Si8Cr5Ni2 particles, the dendritic structures are clearly enriched in Fe. However, in this case, Cr does not seem to mirror the distribution of Fe perfectly but appears almost uniformly distributed over the entire area analysed. The more widespread presence of Cr justifies the conspicuous appearance of silicon-containing phases such as $Cr_3Ni_5Si_2$ and $Al_{13}Cr_4Si_4$ identified by the XRD spectrum. Silicon is more present in Fe-poor areas and has few scattered pure (or particularly enriched) submicron particles. This confirms the presence of pure Si found in the XRD spectrum.

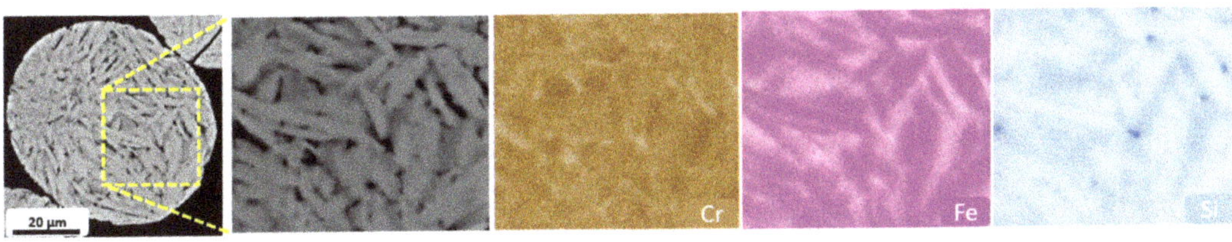

Figure 7. Cross-sectional SEM-EDS maps on cross-sectioned AlFe18Si8Cr5Ni2 powder.

3.4.3. Thermal Analysis by DSC

DSC was used to investigate the precipitation or dissolution reactions of AlFe9Si8Cr2Ni and AlFe18Si8Cr5Ni2 powders. The profile of DSC peaks can, in fact, reveal the occurrence of exothermic or endothermic processes. In particular, the exothermic reaction involves the nucleation and growth of precipitates, releasing energy, while the endothermic reaction corresponds to the dissolution or melting of phases, requiring energy input.

Observing the AlFe9Si8Cr2Ni DSC curve reported in Figure 8, it is possible to note two endothermic peaks between 560 and 620 °C (as onset and offset values). The first peak could be attributable to the melting of the Al-Si eutectic observed by EDS analysis. In fact, the peak temperature (580 °C) is very close to the eutectic temperature of the Al-Si phase diagram (577 °C). Analogously, the second peak could be related to the α-Al melting temperature. In the same temperature range, two endothermic peaks were also detected in the DSC curve of AlFe18Si8Cr5Ni2, as shown in Figure 8. However, here, the peaks appear much less intense, especially the peak relating to the Al-Si eutectic phase, which seems so weak that it reverses the trend and makes the second peak emerge as the most intense. The different nature of the AlFe18Si8Cr5Ni2 DSC peaks could be related to what was observed by EDS and XRD analyses. In particular, XRD and EDS analyses performed on the AlFe18Si8Cr5Ni2 powder revealed that Si tends to form several phases with the other alloying elements, reducing the amount of silicon available for Al. The lower amount of available silicon for Al ideally leads to a shift in the Al-Si phase diagram, resulting in a higher α-Al phase, less liquid available to create the eutectic phase, and a higher α-Al melting temperature. This would explain not only the less intense nature of the peaks in the AlFe18Si8Cr5Ni2 system but also the slight shift to higher temperatures observed in the second peak. The spectra revealed other minor endothermic and exothermic peaks at lower temperatures that may be correlated with the dissolution or precipitation of low-melting phases. However, it is reasonable to assume that reactions related to the higher melting phases, such as Al95Fe4Cr, Cr3Ni5Si2, and Al13Cr4Si4, occur at higher temperatures and are therefore excluded from the analysis performed.

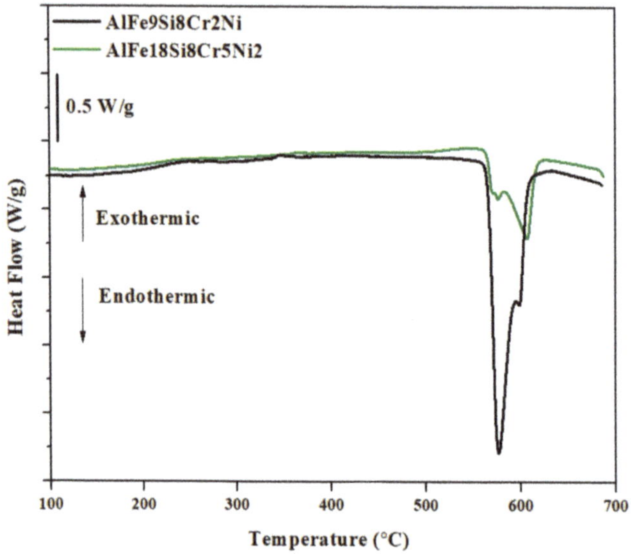

Figure 8. DSC signals for AlFe9Si8Cr2Ni and AlFe18Si8Cr5Ni2 powders.

DSC analyses performed up to 700 °C did not lead to the complete melting of gas-atomised powders. In fact, the powders appeared only partially sintered after DSC analysis. Based on this experimental evidence, it can be asserted that adding 304L to AlSi10Mg led to a significant increase in melting temperature. The study performed by Marola et al. [27] on the AlSi10Mg alloy processed by rapid solidification revealed a melting temperature of below 600 °C. Future analyses at higher temperatures will be conducted to evaluate the reactions related to the higher melting phases detected by XRD and EDS analyses and to quantify the increase in melting temperature in the two systems.

4. Conclusions

This research looked into two previously unexplored gas-atomised Al-Fe-Si-Cr-Ni compositions. To emphasise the importance of sustainability in the development of novel alloys, two widespread alloys in industrial processes, AlSi10Mg and AISI 304L, were used as starting materials for gas atomisation. The selected compositions, AlFe9Si8Cr2Ni and AlFe18Si8Cr5Ni2, were extensively analysed from compositional and microstructural points of view, focusing on alloying element content and particle size. The results obtained are summarised below:

- IGAed compositions exhibited asymmetrical PSD curves, with D50 and D90 percentiles higher than AlSi10Mg, attributed to the presence of Fe, Cr, and Ni altering the molten stream viscosity during gas atomisation.
- AlFe9Si8Cr2Ni and AlFe18Si8Cr5Ni2 particle morphologies were predominantly spherical, with just a limited presence of defects.
- Al-Fe-Si-Cr-Ni powders observed with the BSE detector showed a dendritic pattern with three distinct phases. In AlFe9Si8Cr2Ni, fine particles displayed cellular growth, transitioning to dendritic morphology as particle size increased. In contrast, the AlFe18Si8Cr5Ni2 composition exhibited more elongated and consistent dendritic solidification regardless of particle size.
- In XRD analysis, AlFe9Si8Cr2Ni exhibited intense peaks of fcc-Al, with additional peaks mainly attributed to intermetallic phases $Al_{95}Fe_4Cr$ and $Al_{13}Fe_4$. In AlFe18Si8Cr5Ni2, the same but more pronounced peaks were detected with the addition of peaks related to the presence of the $Cr_3Ni_5Si_2$ and $Al_{13}Cr_4Si_4$ phases.
- EDS mapping of AlFe9Si8Cr2Ni particles showed dendritic structures enriched in Fe and Cr, confirming the $Al_{95}Fe_4Cr$ phase detected with XRD, and interdendritic regions with Al-Si alternations resembling the lamellar structure of AlSi10Mg. In AlFe18Si8Cr5Ni2, dendritic structures appeared to be enriched in Fe and Cr, Al-Si alternations were not detected, and the more uniform Cr distribution justifies the appearance of Cr-containing phases identified by XRD.
- DSC analyses confirmed the higher amount of eutectic Al-Si in the AlFe9Si8Cr2Ni system and revealed a melting temperature of the two compositions well above that of AlSi10Mg.

Author Contributions: Conceptualization, P.F. and M.L.; investigation, A.M., B.B., F.G. and E.V.; data curation, A.M., B.B. and M.L.; writing—original draft preparation, A.M.; writing—review and editing, A.M., B.B., F.B., F.G., D.M. and M.L.; supervision, F.B., D.M., M.L. and P.F. All authors have read and agreed to the published version of the manuscript.

Funding: This research received no external funding.

Data Availability Statement: The data presented in this study are available on request from the corresponding author. The data are not publicly available due to privacy.

Conflicts of Interest: The authors declare no conflict of interest.

References

1. Cáceres, C.H.; Selling, B.I. Casting Defects and the Tensile Properties of an Al-Si-Mg Alloy. *Mater. Sci. Eng. A* **1996**, *220*, 109–116. [CrossRef]
2. Javidani, M.; Larouche, D. Application of Cast Al-Si Alloys in Internal Combustion Engine Components. *Int. Mater. Rev.* **2014**, *59*, 132–158. [CrossRef]
3. Mbuya, T.O.; Odera, B.O.; Ng'ang'a, S.P. Influence of Iron on Castability and Properties of Aluminium Silicon Alloys: Literature Review. *Int. J. Cast Met. Res.* **2003**, *16*, 451–465. [CrossRef]
4. Callegari, B.; Lima, T.N.; Coelho, R.S. The Influence of Alloying Elements on the Microstructure and Properties of Al-Si-Based Casting Alloys: A Review. *Metals* **2023**, *13*, 1174. [CrossRef]
5. Seifeddine, S.; Johansson, S.; Svensson, I.L. The Influence of Cooling Rate and Manganese Content on the β-Al_5FeSi Phase Formation and Mechanical Properties of Al-Si-Based Alloys. *Mater. Sci. Eng. A* **2008**, *490*, 385–390. [CrossRef]
6. Lan, X.; Li, K.; Wang, J.; Yang, M.; Lu, Q.; Du, Y. Developing Al–Fe–Si Alloys with High Thermal Stability through Tuning Fe, Si Contents and Cooling Rates. *Intermetallics* **2022**, *144*, 107505. [CrossRef]
7. Petr, J. The Iron Correctors in Al-Si Alloys. *Int. J. Eng.* **2011**, *9*, 401–405.
8. Aranda, V.A.; Figueroa, I.A.; González, G.; García-Hinojosa, J.A.; Alfonso, I. Study of the Microstructure and Mechanical Properties of Al-Si-Fe with Additions of Chromium by Suction Casting. *J. Alloys Compd.* **2021**, *853*, 157155. [CrossRef]

9. Školáková, A.; Novák, P.; Vojtěch, D.; Kubatík, T.F. Microstructure and Mechanical Properties of Al-Si-Fe-X Alloys. *Mater. Des.* **2016**, *107*, 491–502. [CrossRef]
10. Petrík, J. The Application of Ni for Improvement of Al-Si-Fe Alloys. *Mater. Eng.* **2009**, *16*, 29–32.
11. Xavier, M.G.C.; Freitas, B.J.M.; Koga, G.Y.; Spinelli, J.E. Effects of Ni and Co on the Corrosion Resistance of Al-Si-Cu-Zn-Fe Alloys in NaCl Solution. *Metals* **2022**, *12*, 645. [CrossRef]
12. Lavernia, E.J.; Srivatsan, T.S. The Rapid Solidification Processing of Materials: Science, Principles, Technology, Advances, and Applications. *J. Mater. Sci.* **2010**, *45*, 287–325. [CrossRef]
13. Dunkley, J.J. Metal Powder Atomisation Methods for Modern Manufacturing. *Johns. Matthey Technol. Rev.* **2019**, *63*, 226–232. [CrossRef]
14. Chen, G.; Zhao, S.Y.; Tan, P.; Wang, J.; Xiang, C.S.; Tang, H.P. A Comparative Study of Ti-6Al-4V Powders for Additive Manufacturing by Gas Atomization, Plasma Rotating Electrode Process and Plasma Atomization. *Powder Technol.* **2018**, *333*, 38–46. [CrossRef]
15. Jones, H. On the Prediction of Lattice Parameter vs. Concentration for Solid Solutions Extended by Rapid Quenching from the Melt. *Scr. Metall.* **1983**, *17*, 97–100. [CrossRef]
16. Geoffrey, V.; Raynor, V.G.R. *Phase Equilibria in Iron Ternary Alloys: A Critical Assessment of the Experimental Literature*; Institute of Metals: London, UK; Institute of Metals North American Publications Center: Brookfield, VT, USA, 1988.
17. Ünal, A. Effect of Processing Variables on Particle Size in Gas Atomization of Rapidly Solidified Aluminium Powders. *Mater. Sci. Technol.* **1987**, *3*, 1029–1039. [CrossRef]
18. Prasad, A.; Mosbah, S.; Henein, H.; Gandin, C.A. A Solidification Model for Atomization. *ISIJ Int.* **2009**, *49*, 992–999. [CrossRef]
19. Zhao, Y.; Cui, Y.; Hasebe, Y.; Bian, H.; Yamanaka, K.; Aoyagi, K.; Hagisawa, T.; Chiba, A. Controlling Factors Determining Flowability of Powders for Additive Manufacturing: A Combined Experimental and Simulation Study. *Powder Technol.* **2021**, *393*, 482–493. [CrossRef]
20. Fischmeister, H.F.; Ozerskii, A.D.; Olsson, L. Solidification Structure of Gas-Atomized High-Speed Steel Powders. *Powder Metall.* **1982**, *25*, 1–9. [CrossRef]
21. Xiong, L.; Chuang, A.C.; Thomas, J.; Prost, T.; White, E.; Anderson, I.; Singh, D. Defect and Satellite Characteristics of Additive Manufacturing Metal Powders. *Adv. Powder Technol.* **2022**, *33*, 103486. [CrossRef]
22. Kumar, G.S.; Sateshwar, M.; Sharma, A.R.; Palit, M.; Sarkar, R.; Ghosal, P.; Rao, G.A. Particle Size Dependent Microstructure Evolution of Inert Gas Atomized Nickel Base Superalloy Powders. *J. Alloys Compd.* **2022**, *909*, 164772. [CrossRef]
23. Dai, Y.; Yang, M.; Song, C.; Han, Q.; Zhai, Q. Solidification Structure of $C_{2.08}Cr_{25.43}Si_{1.19}Mn_{0.43}Fe_{70.87}$ Powders Fabricated by High Pressure Gas Atomization. *Mater. Charact.* **2010**, *61*, 116–122. [CrossRef]
24. Fang, P.; Xu, Y.; Li, X.; Chen, Y. Influence of Atomizing Gas and Cooling Rate on Solidification Characterization of Nickel-Based Superalloy Powders. *Xiyou Jinshu Cailiao Yu Gongcheng/Rare Met. Mater. Eng.* **2018**, *47*, 423–430. [CrossRef]
25. Bassini, E.; Galech, U.; Soria, T.; Aristizabal, M.; Iturriza, I.; Biamino, S.; Ugues, D. Effect of the Particle Size Distribution on Physical Properties, Composition, and Quality of Gas Atomized Astroloy Powders for HIP Application. *J. Alloys Compd.* **2022**, *890*, 161631. [CrossRef]
26. Song, D.; Zhao, Y.; Jia, Y.; Li, R.; Zhou, N.; Zheng, K.; Fu, Y.; Zhang, W. Study of the Evolution Mechanisms of Fe-Rich Phases in Al-Si-Fe Alloys with Mn Modification Using Synchrotron X-ray Imaging. *J. Alloys Compd.* **2022**, *915*, 165378. [CrossRef]
27. Marola, S.; Manfredi, D.; Fiore, G.; Poletti, M.G.; Lombardi, M.; Fino, P.; Battezzati, L. A comparison of Selective Laser Melting with bulk rapid solidification of AlSi10Mg alloy. *J. Alloys Compd.* **2018**, *742*, 271–279. [CrossRef]

metals

MDPI

Article

Study on Debinding and Sintering Conditions in Extrusion-Based Additive Manufacturing of 316L and 316L + Cu

Jean-François Silvain [1,2,*], Daniel Lincoln Gifford [1], Sébastien Fourcade [1], Laurent Cuzacq [1], Jean-Luc Grosseau-Poussard [3], Catherine Debiemme-Chouvy [4], Nicolas Tessier Doyen [5] and Yongfeng Lu [2]

1 Bordeaux Institute of Condensed Matter Chemistry, University of Bordeaux, CNRS, Bordeaux INP, ICMCB, UMR 5026, 87 Av. Dr. A. Schweitzer, 33600 Pessac, France; dgifford17@gmail.com (D.L.G.); sebastien.fourcade@icmcb.cnrs.fr (S.F.); laurent.cuzacq@icmcb.cnrs.fr (L.C.)

2 Department of Electrical and Computer Engineering, University of Nebraska-Lincoln, Lincoln, NE 68588-0511, USA; yflu.email@gmail.com

3 LaSie, Pole Science et Technologie, Université de La Rochelle, Av. M. Crépeau, 17042 La Rochelle, France; jlgrouss@univ-lr.fr

4 Laboratoire Interfaces et Systèmes Electrochimiques, UMR 8235 CNRS—SU, Sorbonne Université, CNRS, 4 Place Jussieu, 75005 Paris, France; catherine.debiemme-chouvy@sorbonne-universite.fr

5 IRCER, Institute of Research for Ceramics, IRCER, CNRS, Université de Limoges, CEC, 12 Rue Atlantis, 87068 Limoges, France; nicolas.tessier-doyen@unilim.fr

* Correspondence: jean-francois.silvain@icmcb.cnrs.fr

Abstract: This study investigates the use of a methylcellulose binder in extrusion additive manufacturing of 316L as an alternative to common wax-based binders. Various quantities of copper (Cu) powder were also added in the paste composition to attempt to reduce the sintering temperature by promoting persistent liquid phase sintering. Debinding experiments were conducted under different temperatures and dwell times using argon (Ar), Ar/5%H_2, and Ar/1%O_2 atmospheres. Debinding reduced carbon (C) content to 0.032 wt.% by using a two-step debinding process of Ar/5%H_2 and Ar/1%O_2 thermal treatments. Using this debinding process, sintering was conducted at 1200 °C under Ar/5%H_2 atmosphere with the presence of 0, 10, and 20 vol.% Cu in the paste. Microstructure, mechanical, and corrosion properties were studied. Cu additions allowed the improvement of the densification when sintering at 1200 °C was performed. A 20 vol.% Cu addition yielded 88% relative density after sintering for 10 h, while pure 316L powder sintered under the same conditions had 70%. Mechanical properties were inferior to fully dense stainless steel, but it is not clear if this is due to the Cu additions or insufficient densification.

Keywords: additive manufacturing; 316L; copper; organic paste; mechanical properties

1. Introduction

Additive manufacturing (AM), also known as 3D printing, allows users to rapidly create complex parts without the limitations of traditional shaping and forming processes [1,2]. It uses computer-aided design (CAD) data to build parts with intricate detail and minimal waste. Extrusion-based additive manufacturing (EAM) is a technique that extrudes material through the nozzle of a toolhead onto a platform to form the desired part [3–5]. This technique is seen as a combination of two separate techniques: fused deposition modeling [6] and metal injection molding [7]. Fused deposition modeling (FDM) draws a thermoplastic filament through a heated nozzle using a stepper motor to form polymeric 3D parts. Metal injection molding (MIM) mixes fine metal powder with a polymer binder to form a slurry which is then pumped into a mold to form the part [5,8]. EAM uses the same equipment as FDM but has a toolhead with a syringe attachment which pumps a metal/binder slurry and deposits strands of the material onto a platform layer by layer to form the final part. Interest in this process stems from its cost efficiency when compared to other AM processes. Metallic additive manufacturing techniques often include high-energy arcs and beams and may also require high-temperature environments. For these reasons, companies would be required to invest large capital to apply this technology. Conversely, the EAM process is commonly printed close to room temperature with much lower equipment costs [7,9].

The 316L stainless steel is a common material used for many different applications in medical, construction, automotive and chemical industries. It is valued for maintaining strength and corrosion resistance at elevated temperatures and has shown good performance in a variety of different metal injection molding (MIM) and additive manufacturing (i.e., extrusion additive manufacturing (EAM)) techniques [7,8,10–17].

The nature and composition of a metal–binder slurry is selected based on interactions between the metal and the binder as well as the effectiveness of the debinding process. Most binders are composed of a primary binder and a secondary binder. The primary binder is easily removed at a low temperature and opens up a large network of pores in the part. The secondary binder remains to help maintain the structure of the part until a final thermal treatment. During this treatment, it escapes using the network left by the primary binder by evaporation or reactions with the atmosphere. The binder commonly used with 316L is a mixture of polypropylene, paraffin wax, and stearic acid [8,18,19]. This system is based on the original binder systems developed in the beginning of MIM. Paraffin wax functions as the primary binder. It helps give the paste good flowability and it can be easily removed with a hexane solvent. The stearic acid functions as a surfactant to improve the effectiveness of the paraffin wax and is also removed by hexane. The secondary binder is the polypropylene. It remains in the part after solvent debinding and helps maintain the structure of the part. It is removed by thermal treatment during an intermediate stage of the final sinter [19].

The binder used in this study is a water-based methylcellulose gel. In this case, the primary binder is water. Water assists in the flowability and is then readily removed through a drying process via evaporation. The secondary binder, methylcellulose, is then removed using a thermal treatment. Organic water-based systems are seen as simple and cost-effective. Methylcellulose is also an inexpensive, environmentally friendly, safe compound. Studies have found it to be an effective binder for many different AM applications [20]. Research has not been found using the methylcellulose–316L system, but studies have shown efficacy with the methylcellulose–copper (Cu) system.

To attract companies unequipped for powder metallurgy processing, priority is given to studies that can limit any capital investment. While research has been conducted on printing 316L using EAM and MIM processes [7,8,10], this printing occurs with wax-based binders which involve debinding steps with hazardous hexane or heptane solvents. Traditional sintering temperatures for 316L powder are also high, close to 1400 °C under pure hydrogen atmosphere. This is beyond the capacity of common industrial furnaces, and it requires infrastructure to support a dangerous, flammable gas.

Studying the debinding thermal treatment involves manipulating the temperature, dwell time, heating rate, furnace atmosphere, and gas flow rate. When working with 316L, it is critical to bear in mind that this is a low-carbon (C) austenitic stainless steel with a maximum C content of 0.03 wt.%. For some materials, a small added C content of 0.1 wt.% would be unimportant. With 316L, any C remaining in the material can diffuse and form Cr carbide which will dramatically decrease corrosion resistance. This makes this step particularly important for stainless steel [21].

The sintering temperature for 316L powders for optimal densification is between 1250–1380 °C [22] and is driven by supersolidus liquid phase sintering. The 316L forms a partial liquid phase which assists in the densification of the powder by improving the rate of particle rearrangement and improving the rate of particle distortion through solution–reprecipitation. Conversely, persistent liquid phase sintering is a process that introduces a foreign metal that forms the liquid phase at that metal's melting point [23]. It has the same function as the partial 316L liquid phase, but it allows for sintering at lower temperatures and persists in the matrix after cooling. This process has resulted in large decreases in sintering temperatures for systems like Ni–W and WC–Co [24,25].

Cu has a melting point of 1083 °C; 8–10 wt.% Cu will diffuse into an FCC Fe matrix at 1200 °C, and beyond this it will form a partial liquid phase to assist with sintering. Cu diffuses up to 8–10 wt.% in an Fe matrix, and it would partially dissolve Fe and Cr. It is also completely soluble in nickel (Ni).

While a good deal of research has been conducted on printing 316L using EAM and MIM processes, most studies involve debinding steps with hazardous hexane or heptane solvents. Sintering temperatures for 316L powder are also above 1200 °C and close to 1400 °C under pure hydrogen atmosphere. In order to undergo those specific conditions, this paper studies variations in the post treatment of 316L extruded parts. It tries to eliminate the need for the solvent debinding step by using methylcellulose as an alternative binder. It also explores sintering 316L at a lower temperature of 1200 °C by adding Cu to function as a persistent liquid phase [26]. Lastly, it removes pure hydrogen from the debinding and sintering steps to study using safer gases like argon, Ar/5%H_2, and Ar/1%O_2. Argon has been used as an effective debinding gas in various applications while the Ar/5H_2 functions as a reducing gas for oxides, and Ar/1%O_2 can help fully remove C through oxidation.

2. Materials and Methods

2.1. Creating the Organic Paste and Printing

Obtaining parts by AM by material extrusion requires that the feedstock exhibit appropriate rheological behavior throughout the shaping stage. During extrusion, the material will undergo an intense shear gradient going from zero shear, corresponding to the state of rest in the packaging container, to high shear as it passes through the nozzle and finally shear weak gravity induced on the extruded bead. The rheological properties that govern these transitions are material viscosity, viscoelastic modulus, viscosity recovery behavior, and shear stress. From an experimental point of view, the higher the viscosity of a system, the higher the fidelity during deposition and the better the shape retention properties after extrusion. A material exhibiting ideal behavior must therefore have a low viscosity for shear rates (corresponding to the conditions of the extrusion shaping step) and a sufficiently high viscosity at rest (zero shear) to limit the deformation of the structure. The pastes fabricated in this study have those particular behaviors. Printing parameters can be found in Table S1.

A gas-atomized 316L stainless steel with d_{50} of 12 μm and D_{90} of 23 μm was used for this study. Gas-atomized Cu with d_{50} of 9.5 μm and D_{90} of 18 μm was also used (cf. Figure 1). Both materials were supplied by SAT Nano Technology Material Co., Ltd. Both types of particles had a uniform, spherical shape. Powder composition is reported in Table 1. The methylcellulose used was Methocel® A4M, purchased from Sigma-Aldrich. The primary binder was water with 8 wt.% methylcellulose added to form a gel. This binder was mixed with 45 vol.% 316L particles and Cu amounts varying between 0, 10, and 20 vol.%. Cu and 316L powders were mixed prior to inclusion of the binder. All mixing was performed using a THINKY ARE-250 conditioning mixer (THINKY Corporation, Tokyo, Japan). To evaluate printability of the pastes, printing was conducted using a Lynxter S600D professional 3D printer (Lynxter, Bayonne, France) with a PAS11 toolhead.

Paste compositions were successfully printed into 1 cm^3 cubes, but after evaluation, the samples contained large air pockets. These air pockets were related to the relaxation of the paste and insufficient degassing processes. These air pockets were expected to cause large deviations in densification measurements, so the paste was instead deposited into silicone molds to form bars (length: 10 cm; thickness: 0.90 cm).

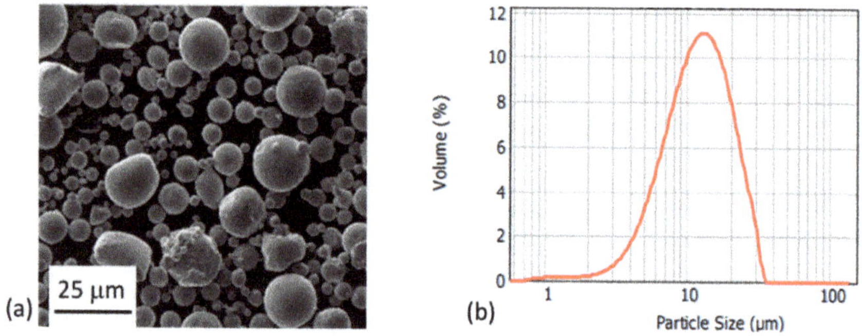

Figure 1. (**a**) SEM micrograph of 316L particles formed by gas-atomization. (**b**) Particle size distribution of those 316L particles.

Table 1. Chemical composition of the 316L powder.

wt.%	Cr	Ni	C	Mn	Mo	Si	P	S	Fe
316L spec.	16–18	10–14	0.03 max.	2.00 max.	2.0–3.0	1.00 max.	0.045 max.	0.030 max.	Bal.
Material used	17.30	12.23	0.017	0.08	2.17	1.00	0.014	0.002	Bal.

2.2. Debinding

The bars were placed in semicylindrical alumina crucibles and debinded using a tubular furnace (Beijing JinYeHong Metallurgical Mechanical Equipment Corp Ltd., Beijing, China). To determine a debinding process sufficient to remove the binder, several different atmospheres, temperatures, and dwell times were evaluated. Argon (Ar), $Ar/5\%H_2$, and $Ar/1\%O_2$ were all tested between 400–700 °C with dwell times from 30 min to 10 h. Temperature heating rates were 3 °C/min. Results showed insufficient carbon (C) removal under argon and $Ar/5\%H_2$ atmospheres across all temperatures and dwell times with retained C measuring from 0.117–0.127 wt.% C.

An $Ar/1\%O_2$ step was added to encourage the removal of C through oxidation. This gas composition was chosen because it is readily available in the industry. There was concern of overoxidation of the material, so temperatures and dwell times were tested to find an optimal quantity to remove the C and minimize 316L oxidation. Oxidation was evaluated using oxygen content measurements. This favored a dwell time of 180 min at 475 °C to yield the lowest oxygen content and still bring the C content to a maximum of 0.032 wt.%.

To minimize the oxygen content in the samples, debinding was first performed under $Ar/5\%H_2$ at 500 °C for 5 h, and then the $Ar/1\%O_2$ step was performed at 475 °C for 3 h.

2.3. Sintering and Hot Rolling

For sintering, the samples were brought to 1200 °C at 3 °C/min under $Ar/5\%H_2$ atmosphere to reduce the oxides on the material before densification. This was performed as a possible nonflammable alternative in place of pure hydrogen. Samples had dwell times of 5 and 10 h. Long dwell times are required to offset the decrease in temperature. Samples were then left to cool in the furnace overnight. Prior to any analysis, samples were mirror-polished on each side. The dimensions of the tested materials (316L, 316L + 10% Cu, and 316l + 20% Cu) were 50 mm × 10 mm and 5 mm in thickness after polishing.

Hot-rolled materials were placed in an oven (Beijing JinYeHong Metallurgical Mechanical Equipment Corp Ltd., Beijing, China), under neutral atmosphere, for 30 min at 500 °C, according to Y. Dong et al. [27], and then immediately hot-rolled (Cavallin 130 mm Rolling Mill (Beijing JinYeHong Metallurgical Mechanical Equipment Corp Ltd., Beijing, China) at a speed close to 10^{-2} m/s. The temperature of the roller close to 100 °C was measured by thermocouple (Prosensor, Amanvillers, France) just prior to hot rolling. Each pass induced a thickness reduction of 10%.

2.4. Methods

Densification measurements were taken using the Archimedes method in water. Samples were mounted and polished to 1 μm using a diamond suspension. Ferric chloride and aqua regia etchants were attempted on samples to reveal the stainless steel particle grain boundaries, but these would overetch the Cu, so were not used. Samples were left unetched and were analyzed using scanning electron microscopy (SEM, Tescan, VEGA © II SBH, Brno, Czech Republic) and energy-dispersive x-ray spectroscopy (EDS) (Bruker, Billerica, MA, USA).

Carbon contents were measured using a CS800 carbon–sulfur determinator (ELTRA®, Carbon analyzer CS-2000, Haan, Germany). Initial carbon content for the 316L powder was 0.0301 wt.% +/− 0.002.

Hardness values are averages of 10 measurements determined by indenting a pyramidal diamond piece with force of 5 kgf (49 N) during 15 s, on a polished surface (WILSON Hardness, Vickers 452 SVD, London, UK).

Mechanical properties of the materials were determined by the ultrasound method (homemade apparatus) This method makes it possible to measure the speed of propagation of longitudinal v_L and transverse v_T waves in a homogeneous, isotropic, and nondispersive material with density ρ. The measurement is carried out by transmission and requires a transmitter and a receiver including a longitudinal wave probe and a transverse wave probe. They will propagate an ultrasonic wave from its surface and determine the speeds v_L and v_T according to the ratio between the thickness e of the material and the transit time t of the ultrasonic wave [28]. The ultrasonic attenuation in the 316L and 316L + Cu samples was measured by pulsed-echo ultrasonic method as per ASTM E664/E664M-10: standard practice for the measurement of the apparent attenuation of longitudinal ultrasonic waves work [29]. A schematic representation of the ultrasonic method is shown in Figure S1.

To compare corrosion resistance, samples were placed in a 5% HCl aqueous solution for 84 h and a corrosion rate was measured.

3. Results

3.1. Material Deposition

Paste compositions were chosen through investigation. The 316L metal content had a larger impact on the printability of the paste than the Cu content. The methylcellulose–316L paste had a uniquely thick, sticky texture. The 316L content was held constant and the total metal loading was adjusted so that all compositions could be printed. The 45 vol.% 316L showed the best results and allowed sufficient printability for the deposition of all the Cu contents. For these compositions, the 316L and 316L + Cu pastes showed a shear thinning behavior which is suitable for an extrusion shaping process.

Pastes were printed into 1 cm^3 cubes formed with uniform extruded pipes, as shown in Figure 2a. One consequence of the sticky paste formulation is the prevalence of air bubbles within the paste. Degassing procedures were unsuccessful in removing the air, causing gaps in the paste flow. After the samples dried, cross-sections also showed air pockets within the material, as seen in Figure 2b. These air pockets were found in gaps between the extruded pipes and a large air pocket was found in the center of every sample. The mechanism for this air pocket is unknown.

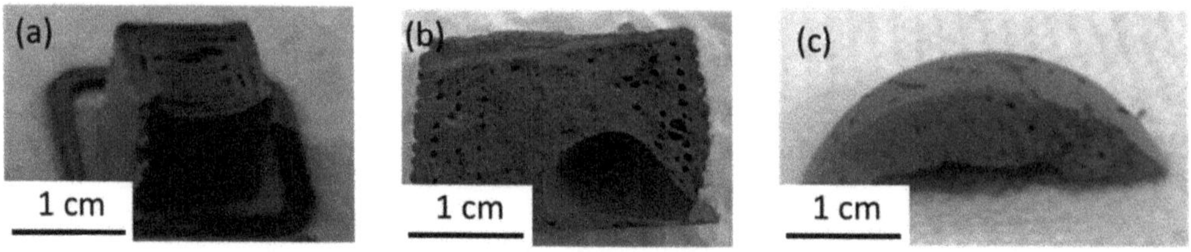

Figure 2. Photographs of (**a**) as-printed sample, (**b**) cross-section of as-printed sample, and (**c**) cross-section of sample formed using a cylindrical mold.

To limit these air pockets, the calendaring process, using silicone molds, provided an alternative that did not result in large gaps within the material. A proper degassing procedure was not successful in preventing air bubbles, so even material formed with molds (cf. Figure 2c) had air pockets, but these were smaller than the printed material. The error caused by these air pockets on the final densification was not quantified. The paste formulations used in this study are extrudable and can be printed into a desired shape. A degassing procedure still needs to be defined and the relaxation of the paste needs to be improved to decrease the initial porosity.

3.2. Debinding

Initial debinding trials were conducted on 316L samples with no Cu addition. Debinding was initially tested under argon and Ar/5%H_2 atmospheres. Thermogravimetric analysis shows that methylcellulose fully degrades between 450–500 °C (cf. Figure S2). Temperatures tested were 400, 500, and 700 °C, with dwell times varying from 90 to 900 min. Heating rates were 2 °C/min and samples were cooled in the furnace overnight. Results are shown in Table 2. The lowest C content was 0.117 wt.%, achieved when the debinding was conducted at 700 °C in Ar/5%H_2 for 300 min. During the C content determination, error for these values were seen as high as +/− 0.002 wt.%, so conclusions on trends cannot be made without more trials.

Table 2. C contents after debinding under various atmospheres, temperatures, and dwell times.

Gas Composition		Temperature	Dwell Time	Carbon Content
Argon (%)	H_2 (%)	(°C)	(min)	(wt.%)
	316L powder (as received)			0.03
100	0.0	400	90	0.130
100	0.0	500	90	0.127
100	0.0	500	300	0.125
95	5.0	500	300	0.125
95	5.0	500	900	0.123
100	0.0	700	300	0.122
95	5.0	700	300	0.117

For debinding at 500 °C, C soot was visible on the surface. Debinding at 500 °C is sufficient to degrade the methylcellulose, but this atmosphere does not effectively remove the C. While 700 °C showed the best C removal, the material had a hard, dark surface with no visible soot or debris. Backscattering micrographs show dark particles and some oxide growth on the sample treated at 700 °C. EDS analysis also shows the presence of a thick Cr oxide on the surface. This oxide may be formed by the reaction of the Cr with the oxygen of the methylcellulose during degradation. The higher temperature promotes this reaction. It is also suspected that the hardness of the samples is related to C diffusion within the sample, causing the formation of Cr carbides. This was not confirmed, but it would negatively impact the corrosion resistance, and would make C removal more difficult. To prevent these things from occurring, the temperature for subsequent thermal treatments was kept at 500 °C.

Large increases in dwell time may result in improved C removal, but excessively long dwell times become impractical for industrial application. Instead, the decision was made to use an oxidizing atmosphere at a lower temperature to remove the remaining C. Heat treatments were conducted at a constant dwell time for different temperatures. The amount of oxidation was measured by determining oxygen contents after treatment. The C content was also measured to determine the amount of removed C. The samples were placed under Ar/1%O_2 for 180 min. Results for different temperatures are shown in Figure 3.

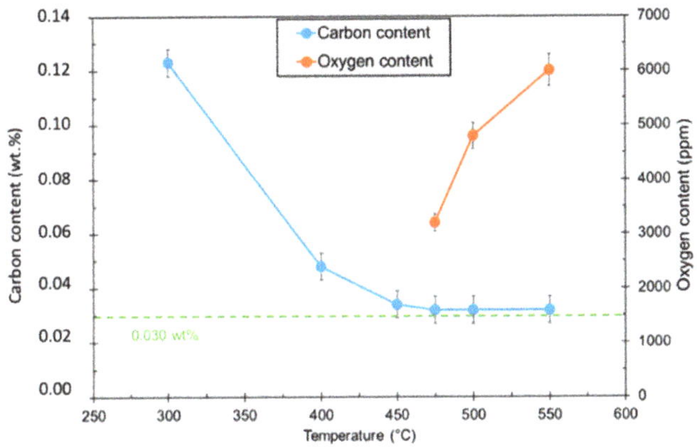

Figure 3. Carbon and oxygen contents of 316L materials after Ar/1%O_2 debinding stage.

C contents show an asymptotic trend, with the C content leveling off at 0.032 wt.% C. For this reason, the temperature chosen for the debinding process was 475 °C. This resulted in minimal oxidation while still achieving the maximum observed C removal.

3.3. Sintering

After debinding, the sintering of the samples with different Cu contents was studied. This was carried out through studying the microstructure, the mechanical properties, and the corrosion properties of the samples. Samples had visual appearances which varied dramatically with the Cu content, as shown in the Supplementary Material Figure S3. All samples containing Cu showed an area at the bottom of the samples where some liquid Cu melted and pooled. All the samples also had a dark green surface on the top and the sides. XRD conducted on these samples showed a large quantity of Cr and silicon oxide on the surface which corresponded to a green color. Only 1 wt.% silicon exists in 316L, but silicon oxide will readily form under the oxidizing conditions seen in the Ar/1%O_2 debinding step. One theory for the oxidation was that oxygen from reduced Fe or Cu oxides reacted with Cr to form Cr oxide. Oxygen contents were then measured on samples before and after sintering to compare values.

These numbers showed a large increase in oxygen contents within the sintered samples.

3.4. Densification

Figure 4 shows the relative density as a function of the sintering dwell time and Table 3 shows the corresponding densities. Traditional analyses of densification compare initial and final densities of the different process variables. In this case, the density prior to sintering, or "green" density, was estimated based on the volume fractions of the material and the material's mass. The binder mostly made up of water had only a small methylcellulose content. The green open porosity was also unable to be directly measured because in this system it was effectively 100%. Many traditional wax-based binders fill gaps between the particles and generate more closed porosity, while low-content binders result in minimal closed porosity. Because of this, direct comparisons before and after sintering are difficult to conduct.

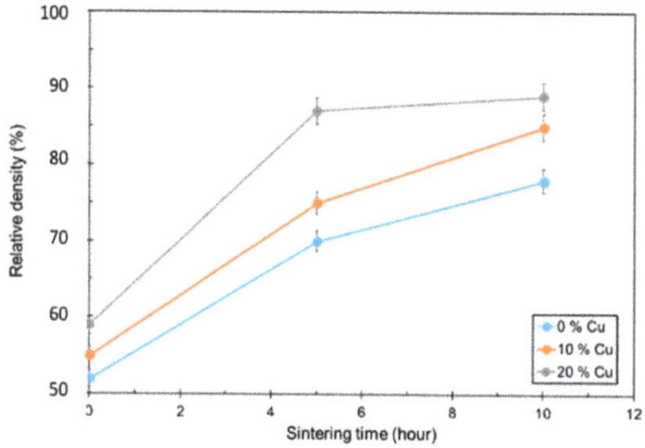

Figure 4. Relative density as a function of sintering dwell time for 0, 10, and 20 vol.% Cu.

Table 3. Relative density and density of 316L, 316L + 10% Cu, and 316L + 20% Cu samples for three annealing times (0 h, 5 h, and 10 h).

		316L	**316L + 10% Cu**	**316L + 20% Cu**
0 h	Relative density (%)	52.17	56.10	59.57
	Density (g/cm^3)	4.12	4.43	4.70
5 h	Relative density (%)	70.40	75.65	87.40
	Density (g/cm^3)	5.36	6.05	6.99
10 h	Relative density (%)	80.00	86.10	88.70
	Density (g/cm^3)	6.49	6.98	7.19

Larger Cu contents resulted in higher densification numbers. The 20 vol.% Cu samples were able to achieve 88% densification after sintering for 5 h. The other samples were approaching similar values as the dwell time increased. The 20 vol.% Cu samples also had the lowest amount of open porosity, and this value did not change with longer dwell times.

When looking at the densification rate as a function of dwell time, one can break down the mechanisms of liquid phase sintering. Density increases dramatically due to the rearrangement of atoms and the introduction of the liquid, until eventually leveling off as the process approaches solid-phase sintering. In this case, larger quantities of copper promote a faster rearrangement step and result in a larger increase in density immediately after liquid formation. The 20 vol.% copper also reaches the solid-phase sintering stage where the density is no longer changing dramatically. The 0 and 10 vol.% copper samples have not leveled off yet, so it is supposed that a longer dwell time would contribute to higher densification numbers.

3.5. *Microstructure*

Etching the samples resulted in preferential overetching of the Cu, so unetched samples were observed under backscattering electron analysis. Figure 5a,c show micrographs of 10 and 20 vol.% Cu samples. Pores are primarily seen within the Cu region instead of within the 316L particles. The primary material seen within the pore sites is silicon oxides, which are highlighted in pink (Figure 5d). It should also be noted that no Cr oxide was observed within the matrix. EDS shows minimal Cu diffusion into the 316L particles, much less than the 8–10 wt.% expected in the Fe–Cu binary phase diagram (Figure 5b,d). Some Ni and molybdenum were also seen to diffuse into the Cu region.

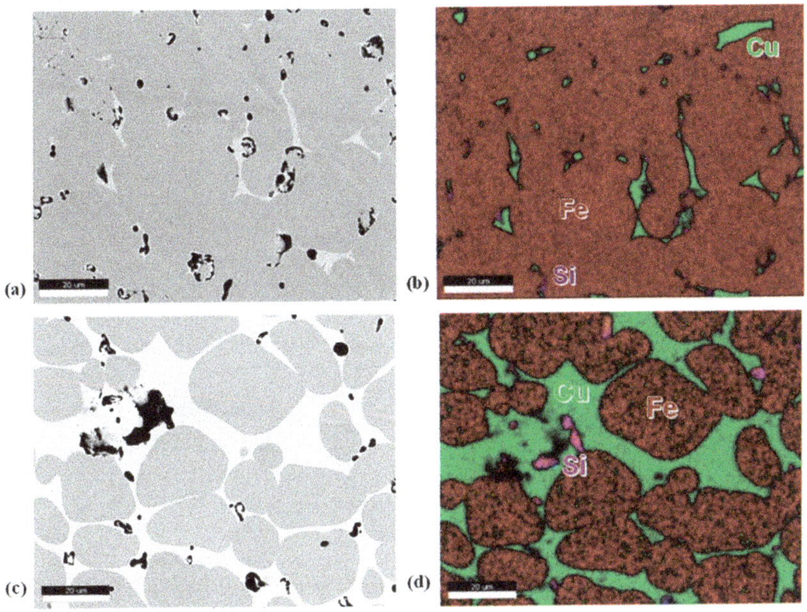

Figure 5. (**a**) 10 vol.% Cu BSE micrograph, (**b**) 10 vol.% Cu EDS map, (**c**) 20 vol.% Cu BSE micrograph, (**d**) 20 vol.% Cu EDS map. For EDS maps, Fe is red, Cu is green, and silicon is pink.

When comparing the 20 vol.% Cu to 0 vol.%, the SEM micrographs in Figure 6 show larger 316L particles in the 20 vol.% Cu material (30% to 50% increase in average size). When referring to the mechanisms for liquid phase sintering, this shows that the Cu is successful in assisting not just the rearrangement of the 316L particles, but particle growth as well.

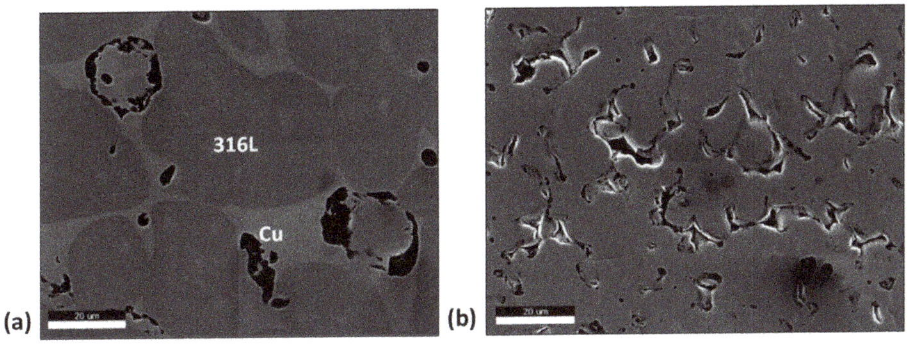

Figure 6. SEM micrographs for (**a**) 20 vol.% Cu, (**b**) 0 vol.% Cu.

The XRD diffractogram shows the presence of a single phase with peaks corresponding to the austenite phase (cf. Figure S4). The initial 316L powder showed mostly the FCC austenite phase with some BCC delta ferrite phase. Different quantities of the alloying elements promoted either the FCC or BCC phase.

The addition of Cu promoted the FCC phase, and after treatment, no BCC phase was seen in the material. The XRD diffractogram shows a clear FCC structure. When Ni diffuses into Cu, the unit cell parameter for the FCC Cu phase is similar to the austenite unit cell, so it only appears as a single phase within the spectrum.

3.6. Mechanical Properties

Vickers hardness results are shown in Figure 7. They show an increase with increasing Cu content that levels off at 20 vol.% Cu and an HV of about 92. Because these values correspond well with the densification values, it was theorized that the increased hardness values are simply due to changes in the densification more than the presence or absence of Cu. To investigate this theory, samples with 20 vol.% Cu were hot-rolled to mechanically reduce the porosity and then subsequently heat-treated. The densification of these samples increased to 94% and the resulting hardness was 150 HV. This shows that improvement of densification will result in a large improvement of mechanical properties. It also shows that the densification will need to be improved to properly compare the effects of Cu additions to standard stainless steel. With this current process, the hardness values fall short of the 316L ASTM standard of 220 HV.

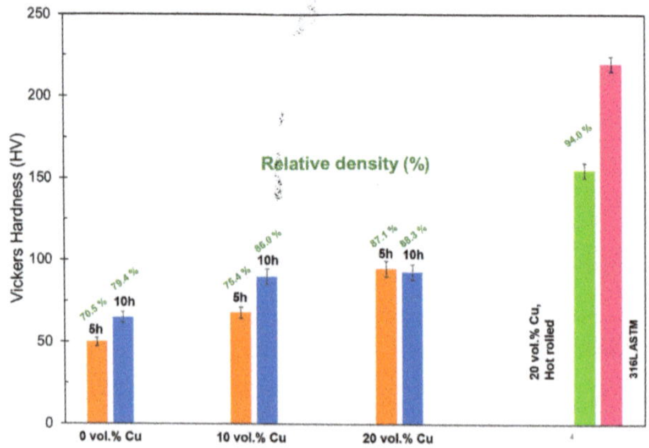

Figure 7. Hardness values for different Cu compositions, after sintering for 5 and 10 h.

The yield strength for the material was estimated using an ultrasonic method and compared to a sample of pure 316L. Results are shown in Figure 8. They show a similar trend with the highest yield strengths found for the largest Cu contents. They also show a consistent decrease in yield strengths for the longer sintering dwell times. These are closer, but still fall short of the minimum 170 MPa requirement for 316L.

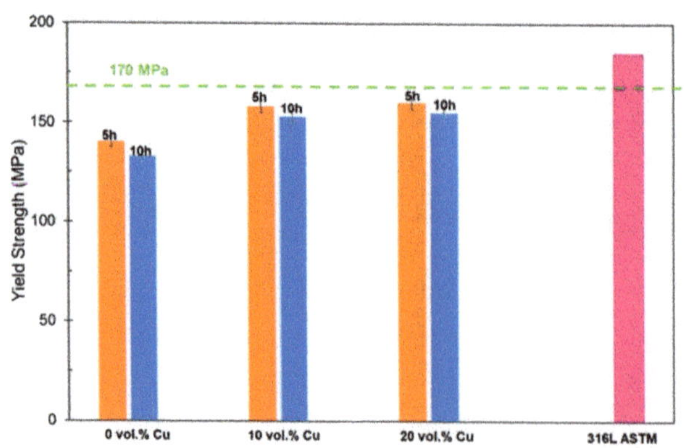

Figure 8. Yield strengths for different compositions for two sintering durations.

3.7. Corrosion Properties

There are many different mechanisms for different types of corrosion in stainless steel. The most common environment studied is using a chloride-bearing aqueous solution similar to seawater. The 316L, because of its chromium (Cr) and molybdenum (Mo) contents, is often a desired choice for seawater applications. Even so, the seawater environment will still initiate pitting and crevice corrosion in 316L. These can be difficult to detect because this type of corrosion creates large cavities which are deep within the material and are only visible as small pits on the material surface. Chlorine ions lead to oxidation of the solid iron (Fe) and form an anodic site surrounded by normal material which functions as a cathode (O_2 reduction).

Corrosion of the 316L samples was evaluated by exposing them to a chloride-bearing environment. Samples were placed in a 5% HCl aqueous solution, and masses were measured gradually over the course of 84 h. This is a process that was performed in a study by He et al. [30], and the results were then compared to the results found in that study. That study achieved 93 and 98% densification. The corrosion rate of the material was then calculated using the following formula:

$$\text{Corrosion rate} = \text{Mass loss}/(\text{Sample area} \times \text{Time}) \quad (1)$$

Results are shown in Table S2. The pure 316L material showed the highest corrosion rate. Flaking was seen on these samples during the corrosion testing which resulted in large mass drops. The corrosion rate was the lowest for the largest Cu content, i.e., 20 vol.%.

4. Discussion

Sintering is conducted under a reducing atmosphere with argon and hydrogen. The reduction of oxides will depend both on the temperature and on the amount of available oxygen. With stainless steel sintering, this is quantified by the dew point. The dew point is related to the ratio of hydrogen to water. A curve summarizing the reduction of Si and Cr oxides is shown in Figure S5. This shows two different dew points resulting in oxide reduction at very different temperatures. Proper sintering of 316L is traditionally performed above 1000 °C with a pure hydrogen atmosphere and a dew point of less than −40 °C [24]. At 1200 °C, Figure S5 shows that the dew point must be less than −20 °C to reduce Cr. Based on gas tank specifications, the inlet gas has a theoretical dew point of −38 °C, but this value has not been confirmed within the furnace itself. When the material is heated to 1200 °C, three major steps occur. The Cu melts at 1083 °C, the Cr oxide reduces, and the silicon oxide reduces. Ideally, both oxides will reduce before melting of the Cu. This would limit the amount of oxygen within the matrix when the sintering starts. At a dew point of −38 °C, the Cr reduces, but it is unclear if the silicon does. After sintering, the material is slowly cooled to room temperature. As the temperature passes below the oxidation temperatures, the Cr and Si on the surface begin to oxidize, and after a long cooling time, the result is a large oxide layer seen on the final part.

Within the sample, the EDS analyses showed that the pores contain large quantities of silicon oxide and no Cr oxide at all. This confirms that the Cr oxide reduces, and any oxygen retained in pores forms silicon oxide. The silicon oxide contents are not seen uniformly in all the pores, so it is not entirely clear if the silicon is reducing at all. Either way, the amount of oxide found in the Cu liquid before reducing will harm the densification of the final material. If the dew point is less than −38 °C, this may affect this quantity. Drops in dew point could be caused by leaking within the furnace, inadequate drying or purging before processing, or higher oxygen contents in the inlet gas. Ultimately, silicon oxide is also undesirable because silicon on the surface is shown to promote pitting corrosion in 316L. To ensure these oxides are reduced, methods must be pursued to manage the dew point and to minimize oxidation after sintering through a faster cooling process. The literature also shows that low cooling allows greater growth of Cr carbides, which has an additional negative effect on corrosion, although it must be noted that this conclusion has not been confirmed with analysis. Because of this effect, most stainless-steel processes require the use of a water or salt quench as the preferred cooling method.

It is hard to compare our results obtained by AEM using bio-sourced paste with 316L materials from the literature due to the fact that base materials, and debinding and sintering conditions are different. Table 4 shows a comparison of 316L materials fabricated using Direct Laser Deposition (DLD), SLM, MIM and EAM with our materials. It has to be mentioned that literature materials are pure 316L where Cu can be added to our 316L in order to improve densification and that 316L materials from the literature are sinter at high temperature (close to 1400 °C) where our materials are sinter at much lower temperature (1200 °C).

Table 4. Overview of mechanical properties and porosity for 316L and 316L + Cu samples produced using fused filament fabrication (FFF), SLM, and MIM.

Porosity (%)	Hardness	Yield Strength	Method	Reference
<0.5	240 HV 380 HV	2200 MPa 2300 MPa	DLD SLM	[31]
-	213 HB	375 MPa	SLM EOS	[32]
-	217 HB	533 MPa	SLM Renishaw	[32]
10	126 HB	153 MPa	FFF BASF	[32]
4	-	170 MPa	MIM	[33]
1.6	200 HV	-	EAM	[34]
~21	65	140 MPa	EAM 316L	This work
~14	85	150 MPa	EAM 316L + 10% Cu	
~12	85	150 MPa	EAM 316L + 20% Cu	

4.1. Densification

When looking at the densification rate as a function of dwell time, one can return to the stages of liquid phase sintering, as shown in Figure S4. The density increases immediately due to the rapid rearrangement of atoms with the introduction of the liquid. Particle distortion and atom reprecipitation occur at a slower rate, until the densification of the system levels off. Higher liquid contents show sharper density increases initially due to more liquid assisting the rearrangement phase. The 10 vol.% Cu sample showed a much more gradual leveling off because it was more dependent on the solution reprecipitation phase. The particles required more distortion to reach the same density as the 20 vol.% Cu which increases the required dwell time. The 0 and 10 vol.% Cu samples had not leveled off at 10 h, so it is supposed that a longer dwell time would contribute to higher densification numbers. It is not clear if these numbers would be higher than the 88% found with 20 vol.% Cu, but yield strength measurements suggest that longer dwell times will negatively affect mechanical properties. Traditionally, final densification is directly connected to the temperature and the dwell time. Lower temperatures require longer dwell times. Dwell times that are too long will cause the material to soften; so, to minimize this, the temperature needs to be increased. One sees a similar relationship with Cu contents. Higher Cu contents result in shorter dwell times, but they are leveling off at a low densification.

Strong capillary forces between the liquid Cu and the 316L particles cause the Cu to penetrate in between the particles, leaving pores behind. Some studies on high-density Fe–Cu compacts have shown particularly aggressive penetration along grain boundaries that can cause a geometric swelling of the material. Large initial porosity prevents this swelling and studies have also shown that Cu's aggressive penetration can actually be useful in filling larger pores [28,35,36]. Studies also show that the addition of C in the Fe matrix helps to minimize Cu diffusion and material swelling. The low Cu contents found within the 316L particles supports these studies. Specific mechanisms for explaining the retained porosity cannot be determined currently.

4.2. Mechanical Properties

The hardness tests correlated the densification closely, so conclusions on mechanical behavior from these values were impossible. The yield strength measurements provided more insight. There was a slight increase in the yield strength with the Cu content, and, like the hardness, it is unclear if this was caused by the addition of Cu or the densification. All samples, however, showed a marked decrease in strength at 10 h when compared to 5 h. As the material coarsens, the density does increase, but the material also softens as the particles combine and the average grain size increases. This softening overtakes the densification to negatively impact the mechanical properties. While sufficient densification may be achieved at longer dwell times, this may also produce a soft, weak material.

4.3. Corrosion Properties

When observing corrosion properties, one must comment on the large quantities of Cr and Si oxides seen on the surface of the material. Pitting corrosion is activated when there is an unexpected fracture or penetration of the Cr oxide layer. There is a random nature of these occurrences which makes the mechanism for these fractures not fully clear in the literature, but they do occur on the surface of the 316L and result in this type of corrosion. It has also been shown that silicon oxides increase the rate of pit activation under a ferric chloride corrosion environment [37]. Once the Cr oxide layer is broken, the low Cr content in the underlying metal will result in more rapid corrosion within the material. In addition to this, if customers wish to clean or further process the densified parts for their application, the cleaning or mechanical removal of this layer will unveil a material with decreased corrosion resistance.

When looking at the corrosion rates, several different mechanisms are working at once. Cu is more prone to corrosion in chloride-bearing environments then 316L. For example, a common etchant for pure Cu is ferric chloride salt. At the same time, greater porosity results in exponential increases in surface area. The data show that Cu additions may have a positive effect on chloride corrosion resistance, but the 316L samples are also much more porous than the samples with liquid Cu. This higher surface area will result in higher corrosion rates. If the actual sample surface area is inaccurate, the corrosion rate will be flawed.

It should also be said that the effect of Cu on the final corrosion properties will be different in different acidic environments like nitric or sulfuric acids. In sulfuric environments, for example, Cu alloying has been shown to improve corrosion resistance through the formation of stable Cu sulfides [17]. These sulfides function similarly to Cr oxide to form a stable protective layer on the surface and protect the stainless steel from forming the pits seen in pitting corrosion. Thus, depending on the application of the material, its corrosion resistance must be measured and evaluated differently.

5. Conclusions

This study investigated the use of methylcellulose–water gel as a binder for 316L powder. It studied effective debinding methods and pursued a low-temperature sintering process. The paste formed with this binder had adequate flow characteristics to successfully print parts that retained their structure. These parts had some difficulty with air pockets which forced the use of a different method for the creation of samples.

The binder was then successfully removed prior to sintering by using a two-step thermal treatment process. The first one used an $Ar/5\%H_2$ atmosphere to degrade the methylcellulose to C. The second one was performed under $Ar/1\%O_2$, which oxidized the C to decrease C contents down to 0.032 wt.% C. Values under that could not be achieved due to C diffusion into the 316L. This provides a safer, environmentally friendly alternative to wax-based binders and solvent baths.

Cu was introduced as a possible way to reduce the sintering temperature by using the persistent liquid phase sintering mechanism. The 10 and 20 vol.% Cu samples were studied and compared to pure 316L powder under the same conditions. Of the samples studied, 20 vol.% Cu sintered for 10 h showed the highest densification, with 88% relative density.

The 20 vol.% Cu sample showed the highest hardness and yield strength. A mechanical decrease in porosity showed a large increase in hardness, which indicates that a higher density will have a dramatic effect on mechanical properties. The yield strength decreased with longer dwell times, showing that the pursuit of maximum densification will result in the compromise of the mechanical properties. Caution must be taken in the future to minimize dwell times.

Conclusions on the corrosion properties were unable to be determined. Low densification results in a high surface area which will damage corrosion resistance. It is also likely that the addition of Cu harms corrosion resistance, but no clear trends were seen.

Sintering was conducted under Ar/5%H_2 instead of pure H_2. There was successful reduction of the Cr oxide, but it is unclear if the Si oxide reduced before the Cu melted. Cr oxide was not found within the sample, but silicon oxide was located within the pores of the material. Any unreduced oxide within the liquid can promote the formation of pores upon reduction. These oxides may need to be reduced to obtain adequate densification. Increases in the hydrogen ratio of the Ar/5%H_2 gas may also be sufficient to reduce these oxides. Finding and eliminating sources of moisture in the furnace may minimize these oxides as well. During slow cooling, a large layer of Cr and silicon oxides is formed on the surface. The effect of rapid cooling rates on the oxide layer and the corrosion resistance may be significant and is worth studying.

Excessive pores may also be the result of insufficient liquid amounts. One study showed that a content of 25 vol.% Cu was required to obtain adequate densification properties [16]. It also included an addition of tin, which the study stated improved the mechanical properties. A study of the content and composition of the liquid can be continued.

Adequate densification has not yet been achieved to yield competitive properties. Even so, this first step showed that methylcellulose may be used as an effective binder for 316L. It also showed that after some development and increased densification, Cu additions may be an effective option to apply extrusion additive manufacturing techniques to the creation of competitive 316L parts.

Supplementary Materials: The following are available online at https://www.mdpi.com/article/10.3390/met13111858/s1, Figure S1: Representation of the ultrasonic method; Figure S2: TGA of the methylcellulose under air and Ar/5% H_2 atmosphere; Figure S3: Photographs of 316L + Cu samples with different Cu contents; Figure S4: XRD diffractograms for (a) pure 316L powder, (b) sample before sintering, and (c) sample after sintering for 5 h; Figure S5: Dew points to reduce Cr and Si oxides at different temperatures; Table S1: Printing parameters; Table S2: Measured corrosion rates in 5% HCl of 316L + Cu samples.

Author Contributions: Conceptualization, J.-F.S. and D.L.G.; methodology, J.-F.S., S.F., J.-L.G.-P. and N.T.D.; validation, J.-F.S., D.L.G., C.D.-C. and Y.L.; formal analysis, D.L.G., N.T.D., S.F. and J.-L.G.-P.; investigation, J.-F.S. and D.L.G.; resources, J.-F.S.; writing—original draft preparation, J.-F.S. and D.L.G.; writing—review and editing, J.-F.S., D.L.G., C.D.-C., L.C. and Y.L.; supervision, J.-F.S. and S.F.; project administration, J.-F.S.; funding acquisition, J.-F.S. All authors have read and agreed to the published version of the manuscript.

Funding: This research received no external funding.

Data Availability Statement: All data needed to evaluate the conclusions in the paper are present in the paper. The datasets generated ruing the current study are available from the corresponding author on reasonable request.

Acknowledgments: The Institute of Condensed Matter Chemistry of Bordeaux (ICMCB) and Lynxter for the use of their equipment and financial support.

Conflicts of Interest: The authors declare no conflict of interest.

References

1. Liu, B.; Wang, Y.X.; Lin, Z.W.; Zhang, T. Creating metal parts by Fused Deposition Modeling and Sintering. *Mater. Lett.* **2020**, *263*, 127252. [CrossRef]
2. Spencer, O.; Yusuf, O.; Tofade, T. Additive Manufacturing Technology Development: A Trajectory Towards Industrial Revolution. *Am. J. Mech. Ind. Eng.* **2018**, *3*, 80–90.
3. Galati, M.; Minetola, P. Analysis of Density, Roughness, and Accuracy of the Atomic Diffusion Additive Manufacturing (ADAM) Process for Metal Parts. *Materials* **2019**, *12*, 4122. [CrossRef] [PubMed]
4. Godec, D.; Cano, S.; Holzer, C.; Gonzalez-Gutierrez, J. Optimization of the 3D Printing Parameters for Tensile Properties of Specimens Produced by Fused Filament Fabrication of 17-4PH Stainless Steel. *Materials* **2020**, *13*, 774. [CrossRef]
5. Suwanpreecha, C.; Manonukul, A. A Review on Material Extrusion Additive Manufacturing of Metal and How It Compares with Metal Injection Moulding. *Metals* **2022**, *12*, 429. [CrossRef]
6. Tosto, C.; Tirillò, J.; Sarasini, F.; Cicala, G. Hybrid Metal/Polymer Filaments for Fused Filament Fabrication (FFF) to Print Metal Parts. *Appl. Sci.-Basel* **2021**, *11*, 444. [CrossRef]
7. Hassan, W.; Farid, M.A.; Tosi, A.; Rane, K.; Strano, M. The effect of printing parameters on sintered properties of extrusion-based additively manufactured stainless steel 316L parts. *Int. J. Adv. Manuf. Technol.* **2021**, *114*, 3057–3067. [CrossRef]
8. Banerjee, S.; Joens, C.J. Debinding and sintering of metal injection molding (MIM) components. In *Handbook of Metal Injection Molding*; Woodhead Publishing: Cambridge, UK, 2019; pp. 129–171.
9. Rane, K.; Strano, M. A comprehensive review of extrusion-based additive manufacturing processes for rapid production of metallic and ceramic parts. *Adv. Manuf.* **2019**, *7*, 155–173. [CrossRef]
10. Thompson, Y.; Gonzalez-Gutierrez, J.; Kukla, C.; Felfer, P. Fused filament fabrication, debinding and sintering as a low cost additive manufacturing method of 316L stainless steel. *Addit. Manuf.* **2019**, *30*, 861. [CrossRef]
11. Gong, H.J.; Snelling, D.; Kardel, K.; Carrano, A. Comparison of Stainless Steel 316L Parts Made by FDM- and SLM-Based Additive Manufacturing Processes. *Jom* **2019**, *71*, 880–885. [CrossRef]
12. Kurose, T.; Abe, Y.; Santos, M.V.A.; Kanaya, Y.; Ishigami, A.; Tanaka, S.; Ito, H. Influence of the Layer Directions on the Properties of 316L Stainless Steel Parts Fabricated through Fused Deposition of Metals. *Materials* **2020**, *13*, 2493. [CrossRef] [PubMed]
13. Damon, J.; Dietrich, S.; Gorantla, S.; Popp, U.; Okolo, B.; Schulze, V. Process porosity and mechanical performance of fused filament fabricated 316L stainless steel. *Rapid Prototyp. J.* **2019**, *25*, 1319–1327. [CrossRef]
14. Alkindi, T.; Alyammahi, M.; Susantyoko, R.A.; Atatreh, S. The effect of varying specimens' printing angles to the bed surface on the tensile strength of 3D-printed 17-4PH stainless-steels via metal FFF additive manufacturing. *MRS Commun.* **2021**, *11*, 310–316. [CrossRef]
15. Suwanpreecha, C.; Seensattayawong, P.; Vadhanakovint, V.; Manonukul, A. Influence of Specimen Layout on 17-4PH (AISI 630) Alloys Fabricated by Low-Cost Additive Manufacturing. *Metall. Mater. Trans. A-Phys. Metall. Mater. Sci.* **2021**, *52*, 1999–2009. [CrossRef]
16. Henry, T.C.; Morales, M.A.; Cole, D.P.; Shumeyko, C.M.; Riddick, J.C. Mechanical behavior of 17-4 PH stainless steel processed by atomic diffusion additive manufacturing. *Int. J. Adv. Manuf. Technol.* **2021**, *114*, 2103–2114. [CrossRef]
17. Rosnitschek, T.; Seefeldt, A.; Alber-Laukant, B.; Neumeyer, T.; Altstädt, V.; Tremmel, S. Correlations of Geometry and Infill Degree of Extrusion Additively Manufactured 316L Stainless Steel Components. *Materials* **2021**, *14*, 5173. [CrossRef]
18. German, R.M. Metal powder injection molding (MIM): Key trends and markets. In *Handbook of Metal Injection Molding*; Woodhead Publishing: Cambridge, UK, 2012; pp. 1–25.
19. Quinard, C.; Barriere, T.; Gelin, J.C. Development and property identification of 316L stainless steel feedstock for PIM and μPIM. *Powder Technol.* **2009**, *190*, 123–128. [CrossRef]
20. Dai, L.; Cheng, T.; Duan, C.; Zhao, W.; Zhang, W.P.; Zou, X.J.; Aspler, J.; Ni, Y.H. 3D printing using plant-derived cellulose and its derivatives: A review. *Carbohydr. Polym.* **2019**, *203*, 71–86. [CrossRef]
21. Heaney, D.F.; Mueller, T.W.; Davies, P.A. Mechanical properties of metal injection moulded 316L stainless steel using both prealloy and master alloy techniques. *Powder Metall.* **2004**, *47*, 367–373. [CrossRef]
22. Müller, L. *Powder Metallurgy Data: Chapter 6 Metal Injection Molding*; Springer: Thale, Germany, 2003; Volume 2A1.
23. German, R.; Suri, P.; Park, S. Review: Liquid phase sintering. *J. Mater. Sci.* **2009**, *44*, 1–39. [CrossRef]
24. ASM Committee. *ASM Handbook*; ASM International: Almere, The Netherlands, 1992.
25. Do Kim, Y.; Ryu, S.; Moon, I. Application of Metal Injection Molding to W-Ni-Fe Heavy Alloy. *J. Jpn. Soc. Powder Powder Metall.* **1999**, *46*, 893–897. [CrossRef]
26. Jamil, S.; Chadwick, G. Investigation and analysis of liquid phase sintering of Fe—Cu and Fe–Cu–C compacts. *Powder Metall.* **2013**, *28*, 65–71. [CrossRef]
27. Dong, Y.; Zhang, Z.; Yang, Z.; Zheng, R.; Chen, X. Effect of annealing temperature on the microstructure and mechanical properties of high-pressure torsion-produced 316L stainless steel. *Materials* **2022**, *15*, 181. [CrossRef] [PubMed]
28. Tessier-Doyen, N. Etude Experimentale et Numerique du Comportement Thermomecanique de Materiaux Refractaires Modeles. Ph.D. Thesis, Universite de Limoges, Limoges, France, 2003.

29. Aghaie-Khafri, M.; Honarvar, F.; Zanganeh, S. Characterization of grain size and yield strength in AISI 301 stainless steel using ultrasonic attenaution measurements. *J. Non Destruct. Eval.* **2012**, *31*, 191–196. [CrossRef]
30. He, H.; Li, Y.; Li, D. Effect of sintering temperature and atmosphere on corrosion behavior of MIM 316L steel. *Adv. Mater. Res.* **2011**, *239–242*, 132–136. [CrossRef]
31. Bedmar, J.; Riquelme, A.; Rodrigo, P.; Torres, B.; Rams, J. Comparison of Different Additive Manufacturing Methods for 316L Stainless Steel. *Materials* **2021**, *14*, 6504. [CrossRef]
32. Kedziora, S.; Decker, T.; Museyibov, E.; Morbach, J.; Hohmann, S.; Huwer, A.; Wahl, M. Strength Properties of 316L and 17-4 PH Stainless Steel Produced with Additive Manufacturing. *Materials* **2022**, *15*, 6278. [CrossRef]
33. Zhang, Y.Y.; Feng, E.S.; Mo, W.; Lv, Y.H.; Ma, R.; Ye, S.L.; Wang, X.G.; Yu, P. On the Microstructures and Fatigue Behaviors of 316L Stainless Steel Metal Injection Molded with Gas- and Water-Atomized Powders. *Metals* **2018**, *8*, 893. [CrossRef]
34. Spiller, S.; Kolstad, S.O.; Razavi, N. Fatigue behavior of 316L stainless steel fabricated via Material Extrusion Additive Manufacturing. *Eng. Fract. Mech.* **2023**, *291*, 544. [CrossRef]
35. Lawcock, R.; Davies, D. Effect of carbon on dimensional and microstructural characteristics of Fe–Cu compacts during sintering. *Powder Metall.* **1990**, *33*, 147–150. [CrossRef]
36. Yang, X.; Bai, Y.L.; Xu, M.; Guo, S.J. Effect of Additive Cu-10Sn on Sintering Behavior and Wear Resistance of 316L Stainless Steel. *J. Iron Steel Res. Int.* **2013**, *20*, 84–88. [CrossRef]
37. Kim, S.T.; Park, Y.S. Effect of copper addition on corrosion behavior of high-performance austenitic stainless steel in highly concentrated sulfuric acid solution-Part 1. *Corrosion* **2007**, *63*, 114–126. [CrossRef]

metals MDPI

Article

A Systematic Study on Layer-Level Multi-Material Fabrication of Parts via Laser-Powder Bed Fusion Process

Andrea Angelastro, Paolo Posa *, Vito Errico and Sabina Luisa Campanelli

Department of Mechanics Mathematics and Management, Polytechnic University of Bari, Via Orabona 4, 70125 Bari, Italy; andrea.angelastro@poliba.it (A.A.); vito.errico@poliba.it (V.E.); sabinaluisa.campanelli@poliba.it (S.L.C.)

* Correspondence: paolo.posa@poliba.it

Abstract: In this work, a systematic study was conducted on the fabrication of multi-material components obtained employing Laser-Powder Bed Fusion (L-PBF) technology. The idea of making multi-material components is a winning capability of additive technologies because it allows for the fabrication of Functionally Graded Materials (FGMs) with the customization of parts according to different required properties. This study aims to determine the ability of an inexpensive system, adaptable to the L-PBF machines already on the market, with a powder-spreading technique based on coaters or rollers, to produce parts with continuously variable properties in each layer. Also, the correlation between certain selectable factors in the production design and the result obtained in terms of metallurgical and mechanical properties and chemical composition was investigated. The factors studied were the relative position of the different materials within the powder chamber and the geometry of the equipment designed to produce the cFGMs components. The performed tests involved the use of two materials, a nickel-based superalloy, and a stainless steel, having different chemical, physical, and mechanical properties to obtain gradual property variations in the manufactured samples. Based on the results of post-process characterization obtained via metallographic, chemical, and mechanical analysis, the relative positions of the materials and the geometry of the developed equipment have a limited effect on the sample's manufactured properties. The characteristics of the FGM zone depend on the nature of the employed powders, and its extent coincides with that defined during the design of the divider.

Keywords: additive manufacturing; laser-powder bed fusion; functionally graded materials; multi-material layer level; nickel superalloy; stainless steel

1. Introduction

There is a growing need in the manufacturing sector to develop methods and processes to produce components that meet increasingly specific requirements. From this perspective, additive manufacturing is an innovative and versatile tool to meet these challenges by enabling the fabrication of components with complex geometries in a relatively cost-effective manner [1–3]. Laser-Powder Bed Fusion (L-PBF) is an additive manufacturing process that is advancing in the production of multi-material metallic structures with complex geometries [4,5]. However, the fabrication of multi-material parts using L-PBF is a challenging task because it requires the deposition of two or more powders in a layer-by-layer fashion [6–10]. This can be achieved using machines specifically designed to process powders of different materials while processing the same layer [4,9,11]. However, this solution requires a prohibitive capital investment. An alternative is to manufacture auxiliary equipment for existing L-PBF machines that would allow the different types of powders to be appropriately separated during the machine set-up. Hence, the powder distribution system then determines the mixing of the materials via the movement of a coater or roller [12]. Therefore, by appropriately utilizing the capabilities of L-PBF machines, it is possible to achieve smooth transitions between different materials (between the layers and at the layer level). This result falls within the realm of Functionally Graded Materials (FGMs), which increase the ability to tailor the final properties of components to their function [13]. In addition, many studies have already pointed out the challenges involved in creating multi-materials. One of the main difficulties lies in fabricating components that combine metal alloys that do not mix easily, such as those that can be differentiated by toughness and hardness. This is because the distinct interface between these alloys can cause the formation of weak metal connections, thermal stress buildup, solubility problems, and discrepancies in thermal properties. FGMs are a promising solution to these problems, as they allow a smooth transition of properties, eliminating abrupt changes at the interface [13,14]. Given their considerable capabilities, FGMs are found to be particularly useful in industrial settings (aerospace applications, energy industry, biomechanical applications, automotive industry, marine applications, etc.).

Several materials have been combined and tested to make FGMs [15–18]. For example, parts can be obtained with a lighter core, while the exterior can be made of a different material compatible with the environment in which the part will be used, avoiding abrupt changes in chemical composition and mechanical properties [19]. In aerospace contexts, there are many situations where a gradient transition between materials is required to reduce thermal stresses, for example, in jet engine components that experience high-temperature gradients; in this case, the use of the stainless steel-SS316L/superalloy-Rene88DT combination was decisive [20,21].

However, research of extreme interest related to FGM fabrication is the study and characterization of powders used in the process. The characterization of metal powders used in L-PBF processes is a topic that many researchers have investigated to determine the correlations between the measurable and observable properties of the particles prior to melting and the quality of the manufactured components. Hilzenthaler et al. investigated the characterization of virgin and recycled steel powders via morphological, chemical, mechanical, and electrical analyses [22]. Among other results, they confirmed that a larger powder particle size results in higher flowability, as also verified in previous work [22,23]. The mechanical properties of the components were also related to the powder particle size in the case of 17-4 PH steel. The results show that the best properties are obtained when the particle diameter is in the range of $20 \div 40$ μm [24]. It is likely that the different physical, chemical, and geometric properties of different powders affect the way they interact during the spreading phase from the powder chamber to the building platform.

Several studies, performed with more complex spreading systems not based on a coater or roller, have verified the feasibility of powder bed laser processing on different materials mixed in the same layer and have investigated the metallurgical properties of the resulting parts [25–27]. Demir and Previtali demonstrated the ability to produce FGM (Fe/Al-12Si) components using a prototype L-PBF machine that allows two different powders to be mixed with different percentages, according to the manufacturer's requirements, and spread from the top using hoppers. This is the case for FGM components between layers, i.e., along the Z-axis of the structure [26]. While Wei et al. demonstrated the ability to produce FGM components within the same layer. In this study, 316L/Cu10Sn materials were combined using a prototype machine patented by the same research group. The prototype system consists of a selective multiple powder distribution system (up to six different materials) operating via programmed ultrasonic vibrations combined with a selective powder removal system [27]. The latter is the case for FGM components layer-level.

Regarding systems using the coater or roller for powder bed spreading, some studies have attempted to use numerical simulations to infer the influence of powder particle size and flowability on the built samples [28,29]. However, various results obtained have been irrelevant due to the strong experimental limitations associated with this fabrication methodology, such as contamination problems, as mentioned extensively in the literature [5]. The main objective of this study is to determine the ability of a low-cost system, adaptable to L-PBF machines already on the market, with a coater or roller for powder bed spreading, to produce parts with continuously variable properties in each layer (cFGMs). Also, the correlation between certain selectable factors in the production design and the result obtained in terms of metallurgical and mechanical properties and chemical composition was investigated. Hence, this is the first systematic experimental work on the influence of the mutual position of the different materials within the powder chamber during the distribution of the powder bed via the coater or the roller. Analyses were made to evaluate whether the initial position of the powders affects the formation of the powder bed and, ultimately, the properties of the manufactured components. In addition, the effect of the geometry of the auxiliary equipment, designed to allow the appropriate separation of different types of powders during machine set-up, on the mechanical and metallurgical properties of the samples produced was evaluated. In order to obtain the desired properties of the samples, tests were conducted using two materials—a nickel-based superalloy and a stainless steel—each with different chemical, physical, and mechanical properties. According to the results of post-process characterization obtained via metallographic, chemical, and mechanical analyses, the relative positions of the materials and the geometry of the developed equipment have a limited effect on the sample's manufactured properties.

2. Materials and Methods

2.1. Materials and Equipment

This experimental work was performed using an L-PBF processing machine via Concept Laser (M1 Cusing). The machine is equipped with a Nd:YAG laser source with a wavelength of 1064 nm. The machine uses a rubber coater as a powder-spreading system that moves from the powder chamber to the building platform. The first material selected for experimentation was AISI 316L austenitic steel provided by Mimete Metalpowders (Osnago, Italy). The other alloy was a nickel-based superalloy provided by Praxair (Danbury, CT, USA). Both powders were obtained via a gas atomization manufacturing process and were sieved to select particle sizes between 15 and 45 µm. The diameter of the sieved particles was measured via SEM image analysis to confirm the sieving quality (Figure 1). The chemical composition of the two powders was determined via EDX microanalysis using a scanning electron microscope, the results of which, compatible with the nominal values indicated by the suppliers, are shown in Table 1. The SEM analysis also showed that the powder particles of both materials had a well-defined spherical shape, with few anomalies, such as satellites or elongated particles, that could affect the proper formation of the powder bed (Figure 1). To properly separate the two materials during machine set-up and subsequently achieve their mixing in a well-defined horizontal band on the building platform, specific auxiliary equipment was designed and constructed. This equipment consists mainly of two vertical thin walls inclined at an appropriate angle α to the coater's direction of movement (Figure 2). The angle was chosen to obtain a mixing zone between the two materials with a width of 20 mm. In this zone, the composition of the mixture changed gradually from the first material to the second, according to a theoretically linear law. In this way, a component called Continuous Functionally Graded Material (CFGM) was obtained.

Table 1. Chemical compositions of the two materials' elements from EDX microanalysis.

Nickel Superalloy [wt.%]					AISI 316L [wt.%]				
C	Ni	Cr	Mn	Co	C	Ni	Cr	Mn	Co
0.16	Bal	15.60	-	10.73	0.02	10.10	16.70	1.00	-
Mo	**Al**	**Ti**	**W**	**Fe**	**Mo**	**Al**	**Ti**	**W**	**Fe**
1.20	2.98	5.17	2.11	-	2.04	-	-	-	Bal.

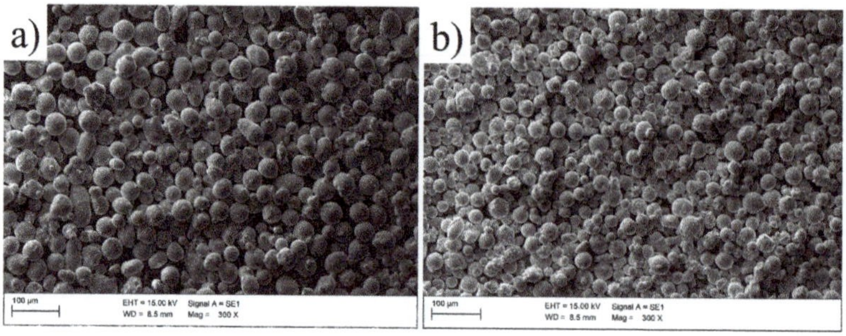

Figure 1. SEM micrograph of the appearance of the powders used in the experiment: (**a**) AISI 316L steel and (**b**) nickel superalloy.

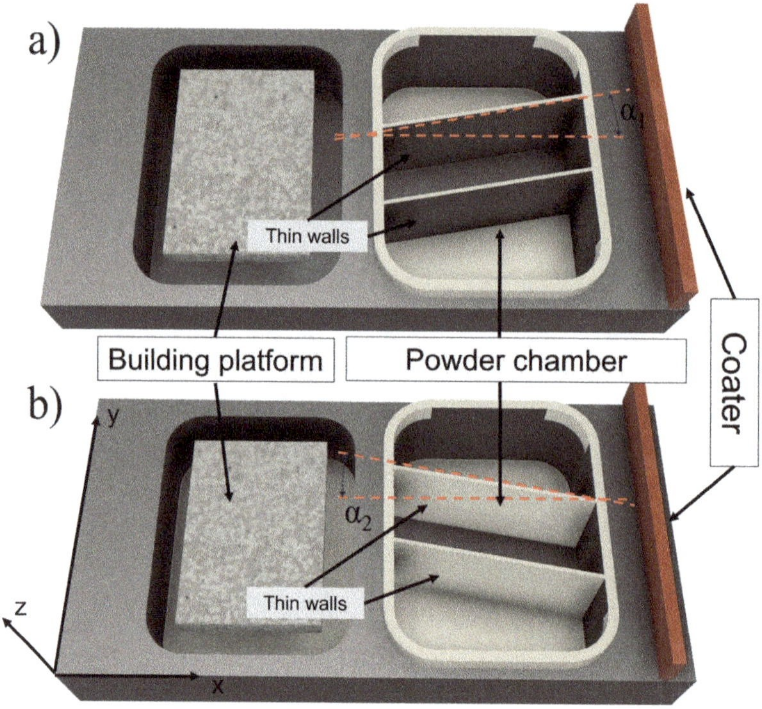

Figure 2. Sketch of the setup used for experimentation showing the two types of dividers: (**a**) α_1-type divider and (**b**) α_2-type divider.

2.2. Experimental Plan and Process Parameters

To maximize the results of each test, two dividers were used, each having two parallel walls. Thus, two different combinations were tested in the same production batch. In fact, in Figure 2a, it can be seen that the walls of the divider were inclined by the angle α_1 with respect to the direction of movement of the coater. This inclination determined the mixing mode between the powder outside the two walls, AISI 316L (henceforth referred to as A), and the powder between them, Nickel Superalloy (henceforth referred to as B). The angle α_1 causes material B to be dragged over material A in the band defined by the top wall of the partition. The mixing mode of the materials is reversed in the zone of influence of the lower wall of the same divider, so that here, material A is dragged onto material B. Figure 2b shows that the walls of the second divider are tilted in the opposite direction to the previous case, resulting in an opposite effect during the distribution and mixing of the materials. The experiment then involved the fabrication of four types of specimens, identified by the angle of wall tilt and the order in which the materials were mixed. The specimens had a total length in the plane (XY) of 30 mm, with a 5 mm zone consisting of AISI316L at one end and a 5 mm zone consisting of the nickel superalloy at the other end. The total height of the specimens along the Z-axis and the length along the X-axis were equal to 10 mm. Between these two zones, there was a transition area 20 mm long, where materials A (AISI316L) and B (nickel superalloy) gradually mixed. A schematic of the composition of the sample is shown in Figure 3.

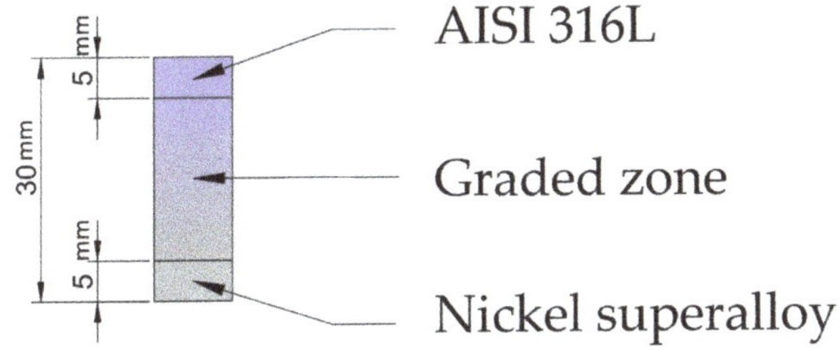

Figure 3. Schematic representation of specimen composition.

For each type of specimen, a replication was made in each print to allow for further investigation. Finally, Figure 4 shows a sketch of the specimens made with the identification used (the symbol " ' " indicates that it is the replication of the specimen with the same name).

Process parameter values optimized in previous work were used to create all samples [12]. Furthermore, the same set of process parameters were used across the whole laser-scanned surface despite the presence of two different materials. The following parameters were used: 100 W laser power (equal to the maximum power of the laser source) and 200 μm laser spot size (not modifiable on the machine used). The layer thickness equal to 30 μm, the hatch spacing equal to 140 μm, and the scanning speed set at 200 mm/s were chosen because it has been demonstrated that this combination of parameters allows porosity in austenitic steel to be reduced to below 1% [30,31]). To minimize the thermal stresses generated during printing, samples were made using the random square island scanning technique, each square with a side of 5 mm.

2.3. Metallographic and Hardness Test

To assess the quality of the fabricated samples qualitatively and quantitatively and to evaluate the influence of the different combinations studied (mentioned in Section 2.2) on the final properties of the produced components, extensive metallurgical (defect, microstructural, and chemical analysis) and mechanical (microhardness tests) characterization was carried out on all the fabricated samples. The metallographic examination was performed by cutting the specimens in a metallographic cut-off machine, hot mounting sections in resin, and polishing them with abrasive papers and diamond pastes. Subsequently, the specimens were observed in the as-polished state to easily quantify the porosity and observe the possible presence of other defects, such as cracks and lack of fusion. A Nikon Eclipse MA200 (Nikon Corporation, Tokyo, Japan) inverted optical microscope was used for this purpose.

The test specimens were subjected to chemical etching to highlight the microstructure and more clearly observe the result of cFGM-oriented fabrication. Glyceregia was used as the chemical reagent. This etching solution consisted of lactic acid, hydrochloric acid, and nitric acid in a ratio of 6:2:1. The analysis of the acquired images was performed using the open-source software ImageJ ver.1.54f, which allowed the extraction of quantitative information that was subsequently processed using Matlab R2023a.

To determine the trends in the composition of the samples, a chemical elemental analysis was performed using a GeminiSEM 500 scanning electron microscope (SEM) (Carl Zeiss Microscopy, New York, NY, USA) equipped with an OXFORD EDS Oxford Aztec Live with Ultim Max 100 detector (Carl Zeiss Microscopy, New York, NY, USA).

The Vickers HV1 mechanical hardness test was performed using Shimadzu HMV-G microhardness tester (Shimadzu, Kyoto, Japan) on the same sections used for the metallographic examination.

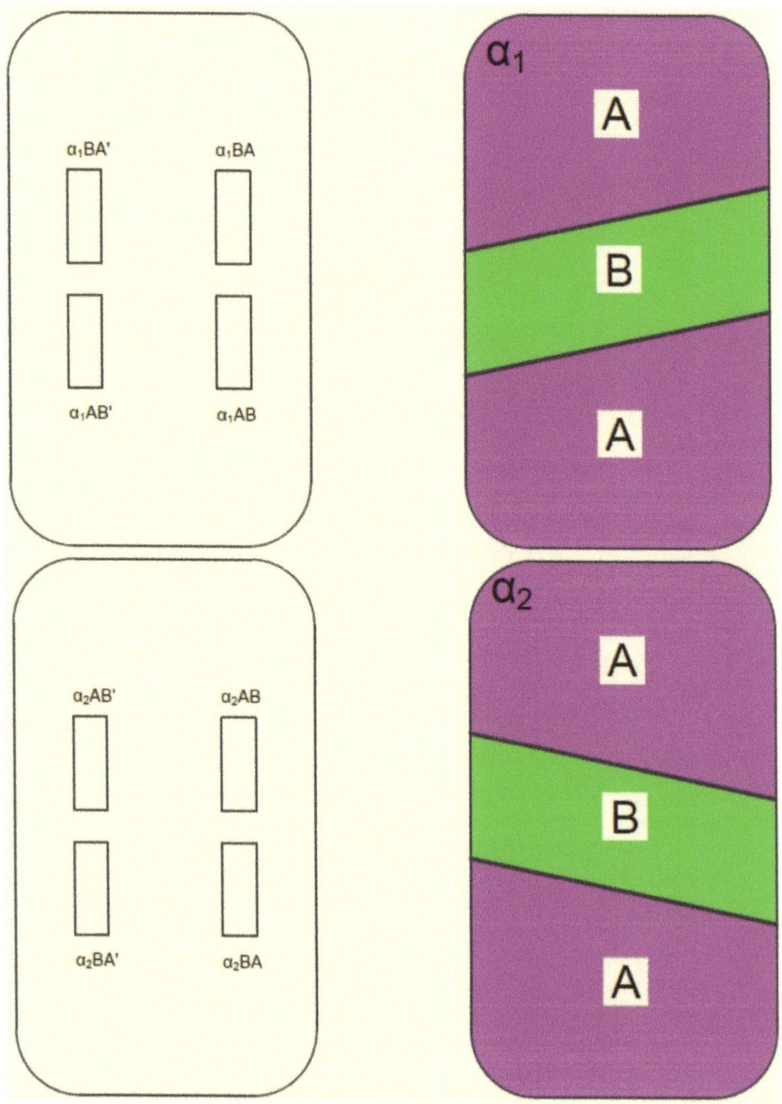

Figure 4. Sketch of the position of the specimens during their manufacture showing the ID assigned to each of them.

3. Results and Discussion

3.1. Metallographic Examination

Metallographic analyses to determine the metallurgical quality and microstructure of the samples were carried out by sectioning in both XY and YZ planes. Considering Figure 4, the identification of the metallographic specimens is shown in Table 2. After polishing the mounted section surfaces, they were observed with an optical microscope and photographed, revealing the appearance of a central band on the specimens that extends along the Y direction (Figure 5). Figure 5 shows that, in all four cases and in the respective replications, the transition zone between the two materials was free of defects such as lack of fusion and delamination, which demonstrated the goodness of the equipment implemented in this study for cFGM fabrication [12].

Table 2. Metallographic specimen identification.

Samples ID	Metallographic Specimen ID	Cut Direction
α_1AB	4T	YZ plane
α_1AB′	4L	XY plane
α_1BA	3T	YZ plane
α_1BA′	3L	XY plane
α_2AB	1T	YZ plane
α_2AB′	1L	XY plane
α_2BA	2T	YZ plane
α_2BA′	2L	XY plane

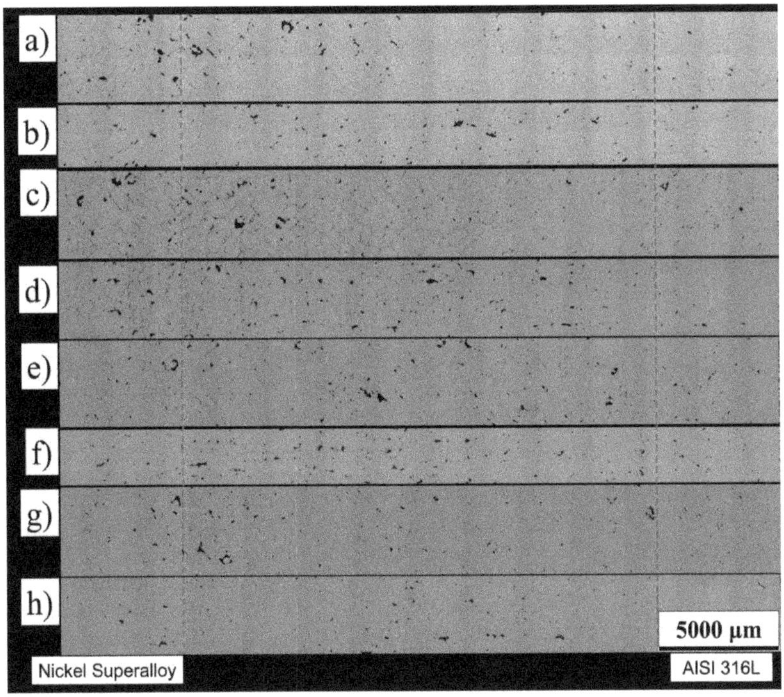

Figure 5. (a–h): Micrographs of the polished surfaces of test specimens 1L, 1T, 2L, 2T, 3L, 3T, 4L, and 4T, respectively.

However, micrographs without chemical etching showed that the two materials behaved differently towards the L-PBF process. The nickel superalloy, which is always on the left in the various images, was affected by numerous cracks that usually plague materials that are not very resistant to rapid melting and resolidification cycles [32]. The austenitic steel, on the other hand, was crack-free and showed only a small percentage of porosity, which is considered physiological in the L-PBF process.

Figure 6 shows the distribution of the number of discontinuities along the XY and YZ planes. The examined sections were divided into bands of 2 mm width, and in each band, the detected discontinuities were counted without taking into account their size. This distribution seemed to follow the same trend as the percentage of nickel superalloy powder, with a maximum value at the left end and tending to decrease, moving toward AISI 316L (see Figure 6a). However, this behavior was not perfectly the same as the angle of the wall changes. In fact, in the metallographic specimens 3 and 4 (see Figure 6c), this tendency is less pronounced, and the total number of discontinuities is also reduced compared with those in specimens 1 and 2 (see Figure 6b).

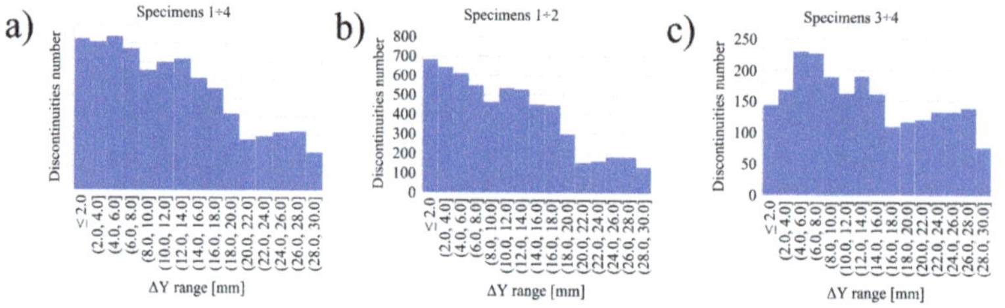

Figure 6. Distribution of the number of discontinuities along the XY and YZ planes: (**a**) distribution in the eight specimens, considering the 4 combinations and (**b**) distribution in specimens 1 and 2 and (**c**) in specimens 3 and 4.

After etching, each area of the specimen composed of one of the two materials showed the typical microstructures of that alloy. In fact, Figure 7a shows an austenitic structure. Furthermore, in the section parallel to the XY plane, the edges of the 45° laser passes are visible, in accordance with the square island strategy used for manufacturing. Figure 7c, on the other hand, shows the microstructure of the nickel superalloy observed in the cross-section, characterized by a dendritic morphology [33]. Figure 7b shows the appearance of the microstructure within the mixed zone, where the characteristics of both materials were found [34].

3.2. Hardness Test

The Vickers HV1 hardness test was carried out according to the international standard ISO 6507-1 [35] using a load application time of 15 s on the surface to be tested. Each specimen listed in Table 2 was tested in six different equispaced zones along the main direction of the specimen, as shown in Figure 8. Three tests were performed at each measurement position, and their mean and standard deviation were calculated to statistically compare the results and verify the influence of different factors on the hardness (see Figures 9 and 10). In particular, an attempt was made to determine how the angle of the divider's wall and the reciprocal position of the powders could affect the mechanical properties of the specimens. However, the comparison between the XY plane and YZ plane sections was not carried out because the L-PBF manufacturing process results in an anisotropic structure in the manufactured components. Therefore, it is obvious to detect differences even in the hardness values measured on perpendicular surfaces. The results showed that although, in all cases, there was a gradual reduction in hardness moving towards AISI 316L, going from ≈370 HV to ≈220 HV, both the angle of the divider and the relative position between the powders had an influence on the evolution of the values. The values obtained were consistent with the principle that the average hardness can be calculated as the sum of the hardness values of the individual phases multiplied by their volume fractions [36].

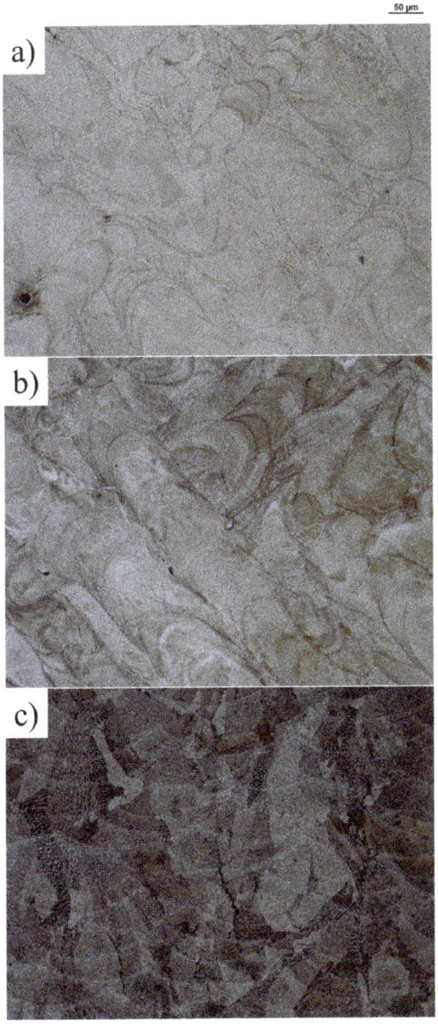

Figure 7. Images of the microstructure of the analyzed samples: (**a**) 4L specimen in the AISI 316L zone; (**b**) 3L specimen in the mixed composition zone; and (**c**) 4T specimen in the nickel superalloy zone.

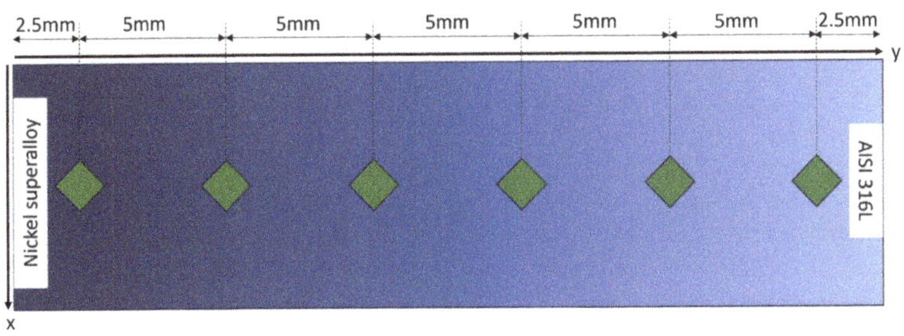

Figure 8. Schematic diagram of hardness measurement positions on the specimen surface.

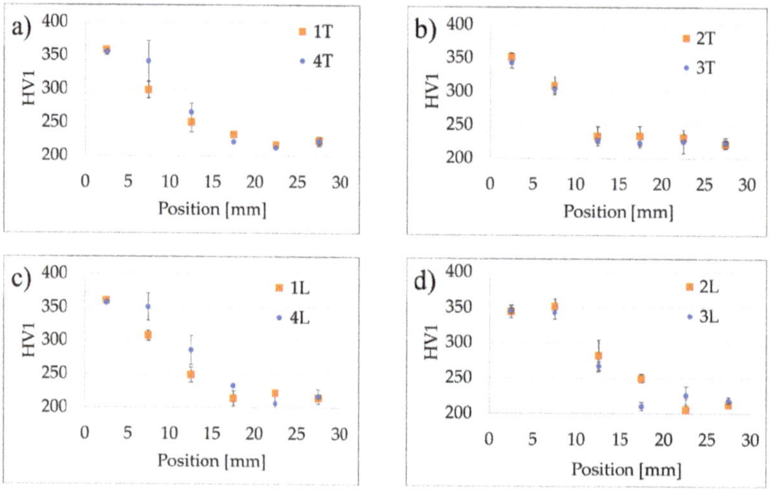

Figure 9. Comparison of hardness values of specimens with different divider angles: (**a**,**c**) hardness results of AB-type specimens and (**b**,**d**) hardness results of BA-type specimens.

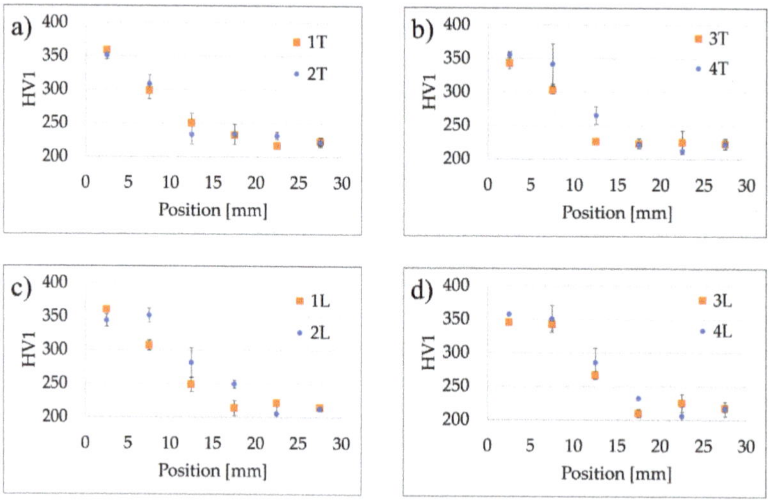

Figure 10. Comparison of hardness values of specimens with different mutual positions of the materials during the distribution of the powder bed: (**a**,**c**) hardness results of α_2 type specimens and (**b**,**d**) hardness results of α_1 type specimens.

Figure 9 shows the summary graphs of the observed hardness trends in the different cases analyzed, comparing the hardness values of the fabricated specimens with the different divider angles. Comparing the graphs, it can be seen that the hardness of the AB-type specimens produced with the angle α_1 tended to remain higher than that of the AB-type specimens produced with the angle α_2 divider up to half the tested length (see Figure 9a,c). Since the higher hardness is due to the greater presence of nickel superalloy, it can be concluded that it was transported to a much greater extent along the Y-direction when the α_1 angle divider was used than in the other case (as for the AB-type samples). This difference is not observed in the case of the BA-type specimens produced with both angles because the values of hardness found were comparable in the different cases (see Figure 9b,d).

Figure 10 shows the summary graphs of the observed hardness trends in the different cases analyzed, comparing the hardness values of the fabricated specimens with different mutual positions of the materials during the distribution of the powder bed. Comparing the graphs, it is possible to show that in the case of the α_2 divider (see Figure 10a,c), the BA-type specimens showed a greater extent of the area affected by the nickel superalloy. This result was due to higher hardness values for almost half of the specimen (a result that was more evident in the case of the XY plane sections). In the case of the α_1 divider (see Figure 10b,d), the same kind of phenomenon is observed in the AB-type specimens (especially in the specimens obtained from the sections along the YZ plane). The results indicate that the factors considered in the experiment have limited influence on the hardness of different samples produced.

3.3. Chemical Analysis

Chemical analysis via scanning electron microscope was carried out on the YZ plane of all samples at 2.5 mm intervals. The weight concentration of the chemical elements differentiating the two materials was measured to see if the theoretical linear variation imposed by the shape of the divider was confirmed via instrumental analysis. The elements present in one material and absent in the other were examined. Specifically, the elements Mn, Fe (present in AISI 316L and absent in the nickel superalloy), Co, Al, and Ti (present in the nickel superalloy and absent in AISI 316L) were analyzed. To better highlight trends in chemical concentrations, the values related to the maximum measured value for each chemical element were plotted. As seen in the graphs in Figure 11, in all four cases studied, all the measured chemical elements showed a weight percentage that follows the expected trend. The first three measurement points, which belonged to the area where only AISI 316L steel was present, showed a flat trend precisely because this was a part of the sample not affected by the cFGM. The same result was found for the last three points, which belonged to the area where only nickel superalloy was present.

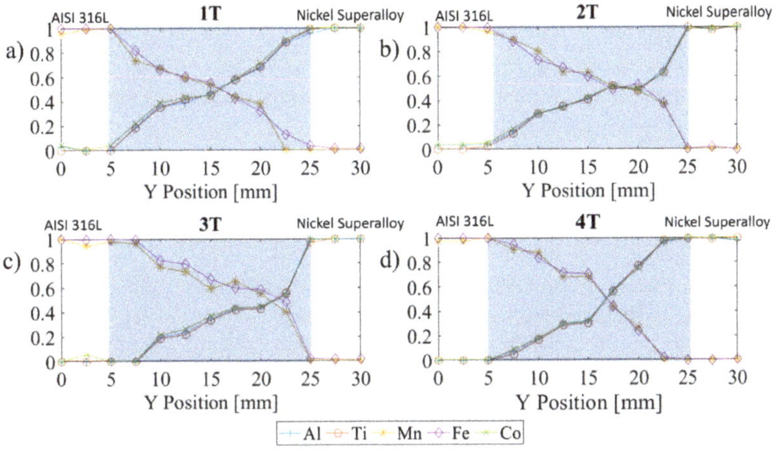

Figure 11. Results of chemical analysis performed on sample: (**a**) 1T, (**b**) 2T, (**c**) 3T, and (**d**) 4T.

These results were identical for all the tested samples. Hence, the designed equipment allowed to obtain components with gradual chemical compositions changing within a predetermined range and in a well-defined bandwidth, avoiding any type of contamination in the other areas. This is an excellent result compared to the results already published in the scientific literature. In fact, as mentioned in the review by Wang et al. [5], L-PBF machines already on the market with a bottom-down powder chamber and a coater-based powder spreading system are severely limited for multi-material fabrication within the same layer due to strong contamination problems.

From the results shown in Figure 11, it can also be noted that the BA-type specimens, in both the α_1 and α_2 cases, showed a less gradual trend; in fact, as seen in both Figure 11b,c, which referred to the 2T and 3T specimens, respectively, there was a sudden change in the chemical concentration of the elements in the 20 mm–25 mm range (along Y-direction). On the other hand, for the AB-type samples, in both cases of α, the change in chemical concentration of the representative elements was more gradual throughout the sample. This result showed that the relative position between materials is influential in cFGM fabrication at the layer level and is a parameter that can be considered, while the different angle of the divider is negligible, at least from the point of view of chemical analysis.

The results of the chemical analysis were consistent with the results of the hardness tests. It can be seen that in the case of the BA-type specimens, a portion of the specimens about 20 mm long had a hardness value very close to that of AISI 316L (see Figure 9b). This result was consistent with what is seen in Figure 11b,c, where about 20 mm of the specimen had a chemical composition very close to that of AISI 316L. The AB-type specimens showed a gradual change both in mechanical properties, as determined via hardness tests (see Figure 9a), and chemical properties (see Figure 11a,d).

4. Conclusions

This experimental work presents a systematic study of the fabrication of multi-material parts at the layer level using the Laser-Powder Bed Fusion process. Specifically, this study aimed to determine the ability of an inexpensive system, adaptable to the L-PBF machines already on the market, to produce parts with continuously variable properties in each layer. Also, the correlation between certain selectable factors in the production design and the result obtained in terms of metallurgical and mechanical properties and chemical composition was investigated. The factors studied were the relative position of the different materials within the powder chamber and the geometry of the equipment designed to produce the cFGMs components. The analyses performed and the main results obtained are summarized below.

- Metallographic examinations have demonstrated the feasibility of producing cFGMs components at the layer level with the system presented in this work; in fact, the interface of the produced samples appeared to be free of defects such as lack of fusion and delamination, synonymous with the good metallurgical bond created between the materials investigated in this work.
- The hardness tests confirmed the presence of a graded zone equal to that set by the divider (20 mm, along the Y direction), within which there was a continuous and gradual change in mechanical properties. In addition, the hardness test results showed a limited influence of both analyzed factors on the final mechanical properties of the produced samples.
- The chemical analyses revealed the variation in chemical properties along the grading direction in all the produced samples; however, from the same analyses, the factor of relative position between different powders was found to have a certain influence on the final result.

According to the results of post-process characterization obtained via metallographic, chemical, and mechanical analyses, it can be stated that the used divider could be designed in different configurations without significantly affecting the properties of the samples produced. The characteristics of the FGM zone depend on the nature of the employed powders, and its extent coincides with that defined during the design of the divider.

Further development of this work could include (i) the study of the tribological characteristics of the powders used and how the contact between chemically and geometrically different particles affects the powder bed coating, also using numerical simulations, and (ii) the study of homogeneity of the FGM zone along the X-axis, using materials with significantly different thermo-physical properties (density, melting temperature, etc.) and particle sizes.

Author Contributions: Conceptualization, V.E. and P.P.; methodology, V.E. and P.P.; software, V.E. and P.P; validation, V.E., A.A. and S.L.C.; formal analysis, V.E. and P.P.; investigation, V.E. and P.P.; resources, A.A. and S.L.C.; data curation, V.E. and P.P; writing—original draft preparation, V.E. and P.P.; writing—review and editing, V.E., A.A. and S.L.C.; visualization, V.E., P.P., A.A. and S.L.C.; supervision, A.A. and S.L.C.; project administration, S.L.C.; funding acquisition, S.L.C. All authors have read and agreed to the published version of the manuscript.

Funding: This research received no external funding.

Conflicts of Interest: The authors declare no conflict of interest.

References

1. Liu, L.; Wang, D.; Deng, G.; Yang, Y.; Chen, J.; Tang, J.; Wang, Y.; Liu, Y.; Yang, X.; Zhang, Y. Interfacial Characteristics and Formation Mechanisms of Copper–Steel Multimaterial Structures Fabricated via Laser Powder Bed Fusion Using Different Building Strategies. *Chin. J. Mech. Eng. Addit. Manuf. Front.* **2022**, *1*, 100045. [CrossRef]
2. Grazia Guerra, M.; Lafirenza, M.; Errico, V.; Angelastro, A. In-Process Dimensional and Geometrical Characterization of Laser-Powder Bed Fusion Lattice Structures through High-Resolution Optical Tomography. *Opt. Laser Technol.* **2023**, *162*, 109252. [CrossRef]
3. Schanz, J.; Islam, N.; Kolb, D.; Harrison, D.K.; De Silva, A.K.M.; Goll, D.; Schneider, G.; Riegel, H. Individual Process Development of Single and Multi-Material Laser Melting in Novel Modular Laser Powder Bed Fusion System. *Prog. Addit. Manuf.* **2022**, *7*, 481–493. [CrossRef]
4. Wei, C.; Zhang, Z.; Cheng, D.; Sun, Z.; Zhu, M.; Li, L. An Overview of Laser-Based Multiple Metallic Material Additive Manufacturing: From Macro- to Micro-Scales. *Int. J. Extrem. Manuf.* **2021**, *3*, 012003. [CrossRef]
5. Wang, D.; Liu, L.; Deng, G.; Deng, C.; Bai, Y.; Yang, Y.; Wu, W.; Chen, J.; Liu, Y.; Wang, Y.; et al. Recent Progress on Additive Manufacturing of Multi-Material Structures with Laser Powder Bed Fusion. *Virtual Phys. Prototyp.* **2022**, *17*, 329–365. [CrossRef]
6. Wei, C.; Li, L. Recent Progress and Scientific Challenges in Multi-Material Additive Manufacturing via Laser-Based Powder Bed Fusion. *Virtual Phys. Prototyp.* **2021**, *16*, 347–371. [CrossRef]
7. Gu, H.; Wei, C.; Li, L.; Han, Q.; Setchi, R.; Ryan, M.; Li, Q. Multi-Physics Modelling of Molten Pool Development and Track Formation in Multi-Track, Multi-Layer and Multi-Material Selective Laser Melting. *Int. J. Heat Mass Transf.* **2020**, *151*, 119458. [CrossRef]
8. Mussatto, A. Research Progress in Multi-Material Laser-Powder Bed Fusion Additive Manufacturing: A Review of the State-of-the-Art Techniques for Depositing Multiple Powders with Spatial Selectivity in a Single Layer. *Results Eng.* **2022**, *16*, 100769. [CrossRef]
9. Wei, C.; Li, L.; Zhang, X.; Chueh, Y.H. 3D Printing of Multiple Metallic Materials via Modified Selective Laser Melting. *CIRP Ann.* **2018**, *67*, 245–248. [CrossRef]
10. Wei, C.; Gu, H.; Sun, Z.; Cheng, D.; Chueh, Y.-H.; Zhang, X.; Huang, Y.; Li, L. Ultrasonic Material Dispensing-Based Selective Laser Melting for 3D Printing of Metallic Components and the Effect of Powder Compression. *Addit. Manuf.* **2019**, *29*, 100818. [CrossRef]
11. Pajonk, A.; Prieto, A.; Blum, U.; Knaack, U. Multi-Material Additive Manufacturing in Architecture and Construction: A Review. *J. Build. Eng.* **2022**, *45*, 103603. [CrossRef]
12. Errico, V.; Posa, P.; Fusco, A.; Angelastro, A.; Campanelli, S.L. Intralayer Multi-Material Structure Stainless-Steel/Nickel-Superalloy Fabricated via Laser-Powder Bed Fusion Process. *Manuf. Lett.* **2023**, *35*, 11–15. [CrossRef]
13. Reichardt, A.; Shapiro, A.A.; Otis, R.; Dillon, R.P.; Borgonia, J.P.; McEnerney, B.W.; Hosemann, P.; Beese, A.M. Advances in Additive Manufacturing of Metal-Based Functionally Graded Materials. *Int. Mater. Rev.* **2021**, *66*, 1–29. [CrossRef]
14. Ye, Z.; Huang, J.; Cheng, Z.; Wang, S.; Yang, J.; Chen, S.; Zhao, X. Interfacial Characteristics and Mechanical Properties of Aluminum/Steel Butt Joints Fabricated by a Newly Developed High-Frequency Electric Cooperated Arc Welding-Brazing Process. *J. Mater. Process. Technol.* **2021**, *298*, 117317. [CrossRef]
15. Dubey, A.; Jaiswal, S.; Lahiri, D. Promises of Functionally Graded Material in Bone Regeneration: Current Trends, Properties, and Challenges. *ACS Biomater. Sci. Eng.* **2022**, *8*, 1001–1027. [CrossRef]
16. Rouf, S.; Malik, A.; Raina, A.; Irfan Ul Haq, M.; Naveed, N.; Zolfagharian, A.; Bodaghi, M. Functionally Graded Additive Manufacturing for Orthopedic Applications. *J. Orthop.* **2022**, *33*, 70–80. [CrossRef]
17. Sam, M.; Jojith, R.; Radhika, N. Progression in Manufacturing of Functionally Graded Materials and Impact of Thermal Treatment—A Critical Review. *J. Manuf. Process.* **2021**, *68*, 1339–1377. [CrossRef]
18. Zhang, B.; Jaiswal, P.; Rai, R.; Nelaturi, S. Additive Manufacturing of Functionally Graded Material Objects: A Review. *J. Comput. Inf. Sci. Eng.* **2018**, *18*, 041002. [CrossRef]
19. Ansari, M.; Jabari, E.; Toyserkani, E. Opportunities and Challenges in Additive Manufacturing of Functionally Graded Metallic Materials via Powder-Fed Laser Directed Energy Deposition: A Review. *J. Mater. Process. Technol.* **2021**, *294*, 117117. [CrossRef]
20. Lin, X.; Yue, T.M.; Yang, H.O.; Huang, W.D. Phase Formation and Microstructure Evolution in Laser Rapid Forming of Graded SS316L/Rene88DT Alloy. *Mater. Sci. Eng. A* **2005**, *402*, 325–336. [CrossRef]

21. Lin, X.; Yue, T.M.; Yang, H.O.; Huang, W.D. Laser Rapid Forming of SS316L/Rene88DT Graded Material. *Mater. Sci. Eng. A* **2005**, *391*, 325–336. [CrossRef]
22. Hilzenthaler, M.; Bifano, L.; Scherm, F.; Fischerauer, G.; Seemann, A.; Glatzel, U. Characterization of Recycled AISI 904L Superaustenitic Steel Powder and Influence on Selective Laser Melted Parts. *Powder Technol.* **2021**, *391*, 57–68. [CrossRef]
23. Sutton, A.T.; Kriewall, C.S.; Karnati, S.; Leu, M.C.; Newkirk, J.W. Characterization of AISI 304L Stainless Steel Powder Recycled in the Laser Powder-Bed Fusion Process. *Addit. Manuf.* **2020**, *32*, 100981. [CrossRef]
24. Sefene, E.M. State-of-the-Art of Selective Laser Melting Process: A Comprehensive Review. *J. Manuf. Syst.* **2022**, *63*, 250–274. [CrossRef]
25. Liu, Z.H.; Zhang, D.Q.; Sing, S.L.; Chua, C.K.; Loh, L.E. Interfacial Characterization of SLM Parts in Multi-Material Processing: Metallurgical Diffusion between 316L Stainless Steel and C18400 Copper Alloy. *Mater. Charact.* **2014**, *94*, 116–125. [CrossRef]
26. Demir, A.G.; Previtali, B. Multi-Material Selective Laser Melting of Fe/Al-12Si Components. *Manuf. Lett.* **2017**, *11*, 8–11. [CrossRef]
27. Wei, C.; Sun, Z.; Chen, Q.; Liu, Z.; Li, L. Additive Manufacturing of Horizontal and 3D Functionally Graded 316L/Cu10Sn Components via Multiple Material Selective Laser Melting. *J. Manuf. Sci. Eng. Trans. ASME* **2019**, *141*, 081014. [CrossRef]
28. Escano, L.I.; Parab, N.D.; Xiong, L.; Guo, Q.; Zhao, C.; Sun, T.; Chen, L. Investigating Powder Spreading Dynamics in Additive Manufacturing Processes by In-Situ High-Speed X-Ray Imaging. *Synchrotron Radiat. News* **2019**, *32*, 9–13. [CrossRef]
29. Habiba, U.; Hebert, R.J. Powder Spreading Mechanism in Laser Powder Bed Fusion Additive Manufacturing: Experiments and Computational Approach Using Discrete Element Method. *Materials* **2023**, *16*, 2824. [CrossRef]
30. Errico, V.; Fusco, A.; Campanelli, S.L. Effect of DED Coating and DED + Laser Scanning on Surface Performance of L-PBF Stainless Steel Parts. *Surf. Coat. Technol.* **2022**, *429*, 127965. [CrossRef]
31. D.'Accardi, E.; Palumbo, D.; Errico, V.; Fusco, A.; Angelastro, A.; Galietti, U. Analysing the Probability of Detection of Shallow Spherical Defects by Means of Pulsed Thermography. *J. Nondestr. Eval.* **2023**, *42*, 27. [CrossRef]
32. Xu, J.; Kontis, P.; Peng, R.L.; Moverare, J. Modelling of Additive Manufacturability of Nickel-Based Superalloys for Laser Powder Bed Fusion. *Acta Mater.* **2022**, *240*, 118307. [CrossRef]
33. Park, J.-U.; Jun, S.-Y.; Lee, B.H.; Jang, J.H.; Lee, B.-S.; Lee, H.-J.; Lee, J.-H.; Hong, H.-U. Alloy Design of Ni-Based Superalloy with High Γ' Volume Fraction Suitable for Additive Manufacturing and Its Deformation Behavior. *Addit. Manuf.* **2022**, *52*, 102680. [CrossRef]
34. Ben-Artzy, A.; Reichardt, A.; Borgonia, J.P.; Dillon, R.P.; McEnerney, B.; Shapiro, A.A.; Hosemann, P. Compositionally Graded SS316 to C300 Maraging Steel Using Additive Manufacturing. *Mater. Des.* **2021**, *201*, 109500. [CrossRef]
35. *ISO 6507-1:2018*; Metallic Materials—Vickers Hardness Test—Part 1: Test Method. ISO: Geneva, Switzerland, 2018.
36. Liyanage, T.; Fisher, G.; Gerlich, A.P. Influence of Alloy Chemistry on Microstructure and Properties in NiCrBSi Overlay Coatings Deposited by Plasma Transferred Arc Welding (PTAW). *Surf. Coat. Technol.* **2010**, *205*, 759–765. [CrossRef]

metals

MDPI

Article

Research on the Microstructure and Mechanical Properties of Repaired 7N01 Aluminum Alloy by Laser-Directed Energy Deposition with Sc Modified Al-Zn-Mg

Jibing Chen [1], Shanji Yu [1], Junsheng Yang [1,*], Rong Xu [2], Ruidi Li [2], Shisen Huang [1], Hongbin Zhu [3] and Xinyan Liu [2]

1 School of Mechanical Engineering, Wuhan Polytechnic University, Wuhan 420023, China
2 State Key Laboratory of Powder Metallurgy, Central South University, Changsha 410083, China
3 CRRC Industrial Research Institute, Beijing 100053, China
* Correspondence: yangjunsheng2008@163.com

Abstract: Aluminum alloy is an important material used in railway train structures. It is of great significance to repair aluminum alloy through directional energy deposition to reduce cost and improve the performance of the aluminum alloy. In this study, 7N01 aluminum alloy was repaired by means of laser-directed energy deposition (DED) with the powder of Sc-modified Al-Zn-Mg aluminum alloy as raw material. The microstructure and mechanical properties of the repaired specimens were studied through the metallographic microscope, scanning electron microscope, electron backscatter diffraction, universal tensile test, and Vickers hardness test in combination. The results show that the bonding interface of the repaired aluminum alloy is satisfactory, and the porosity is 2.8%. The grains in the repaired area are the columnar crystals growing vertically along the boundary of the melt pool with an obvious temperature gradient. Fine equiaxed crystals are distributed along the boundary of the melt pool, and Al_3(Sc,Zr) particles play a role in grain refinement. The average grain size of the fine grain area in the repair zone next to the fusion zone is 9.1 μm, and the average grain size of the coarse grain area is 20 μm. The average tensile strength in the area of repair approaches 349 MPa, which is 91% that of the base material, and the elongation rate is 10.9%, which is 53.2% that of the base material. The hardness ranges between 122 HV and 131 HV, which is comparable to the base material. However, there is a significant decrease in the tensile strength and hardness of the base material (heat-affected zone).

Keywords: Sc modification; DED; microstructure; mechanical property; heat-affected zone

1. Introduction

With various advantages such as high strength, good plasticity, and high performance in corrosion resistance, aluminum alloy can effectively reduce the self-weight of components and maintain a certain level of strength and toughness. Therefore, it plays an important role in promoting the application of large-sized complex aluminum alloys in trains. When the train is operational in a harsh environment for long, the components made of aluminum alloy tend to be damaged (such as wearing, stress corrosion, and fatigue crack), which can lead to failure, thus posing safety risks to the passengers onboard [1]. Therefore, the prompt repair of damaged parts can not only reduce operational costs but also ensure the safety of passengers. With a laser beam as the source of energy, a closed-loop control system is applied to produce the metallic components that can be used to control dimensional accuracy, physical integrity, and full density [2]. Due to the high thermal conductivity and surface reflectivity of aluminum, heat tends to accumulate in the course of processing, thus resulting in cracks and holes. Therefore, high laser power and an appropriate scanning speed are required to melt aluminum alloy powder. With the increase of laser energy, the alloying elements with a low boiling point, such as zinc and magnesium, are selectively evaporated, which may lead to a decrease in the porosity and mechanical properties of the deposited samples [3].

Due to the high coefficient of thermal expansion, high-temperature ranges from liquids to dissolved phase line, and significant shrinkage after solidification, Al-Zn-Mg alloy is prone to cracking when DED solidifies rapidly. The extensive thermal cracking in Al-Zn-Mg alloys produced with laser additives causes deterioration in their properties, even to the extent that they are inferior to conventionally processed alloys. In general, the composition of aluminum alloy determines the development of hot cracks, so optimizing the composition of the alloy can be effective in reducing or preventing the generation of hot cracks [4]. Anika Langebeck et al. used DED to print EN AW-7075 alloy by increasing laser power to extend degassing, which not only improves hardness and strength but also reduces porosity, with the tensile strength reaching up to (222 ± 17) MPa [5]. M.J. Benoit printed Al 7075 alloy by adjusting various parameters of laser processing, such as laser power, scanning speed, and mass flow rate. With the extensive solidification-induced cracks eliminated, the porosity falls below 1% [6]. Wang Xiaoyan et al. repaired 7050 aluminum alloy with $AiSi_{12}$ and effectively inhibited the development of pores and cracks by optimizing process parameters and changing the state of the powder [7].

The DED process is characterized by rapid solidification, and solidification-induced cracking is a problem discovered in 7xxx alloy DED. It has been indicated that introducing a small amount of Sc to 7 series aluminum alloy can provide the nucleation sites required to promote the formation of equiaxed grain $Al_3(Sc,Zr)$, which makes it more resistant to cracking than columnar grain when solidifying, thus improving its mechanical properties [4,8]. Pan Wei et al. used laser selective melting (SLM) to print Al-6.2Zn-2Mg-xSc-xZr alloy and studied the microstructure characteristics of Al-6.2Zn-2Mg-xSc-xZr alloy with different Sc and Zr contents, as well as the influence mechanism of Sc and Zr on cracking. The research results indicate that with the increase of Sc and Zr content, the crack trend and grain size in the printed samples decrease. When the Sc and Zr contents reach 0.6% and 0.36%, respectively, cracks no longer appear in the printed sample [9]. Bi Jiang et al. obtained SLM samples under different process parameters by mixing low microalloy content (0.4 wt% Sc and 0.25 wt% Zr) with Al 7075 powder. According to the results, a higher laser energy density is conducive to reducing microcracks and refining grains, which significantly improves compressive strength [10].

It has been revealed in plenty of studies that the heat-treatable aluminum alloy softens in the heat-affected zone after welding, which is mainly attributed to the "over-aging" of the zone at high welding temperatures. That is to say, the second phase cograted with the aluminum matrix is precipitated, agglomerates, and grows, thus reducing the strengthening effect [11]. Li Nianjun et al. took 7075-T6 aluminum alloy as a plate for friction stir processing, with the research results showing that the mechanical properties (tensile strength and elongation) deteriorated significantly after friction stir processing [12]. The temperature of 7xxx aluminum alloy in the DED process is higher, and a similar situation occurs. At present, there has been a lot of research conducted on the laser welding of 7 series aluminum alloys both at home and abroad [13]. However, there are still few reports on the laser repair of 7 series aluminum alloy, especially DED technology. In the present study, Sc-modified Al-Zn-Mg alloy powder is used to explore the approach to laser additive repair of 7N01 aluminum alloy for an in-depth analysis of the microstructure and mechanical properties of the repaired samples. This is purposed to provide a theoretical reference for the laser repair technology applicable to 7 series aluminum alloy.

2. Materials and Methods

2.1. Experimental Methods and Materials

The base material used in this study is 7N01 (annealed) aluminum alloy. In the base plate sized 150 mm × 100 mm × 11 mm, a groove is opened to simulate the damage. Then, the aluminum alloy base plate to be repaired is slotted, whose cross-section is shown in Figure 1. The surface of the groove is 9 mm in length, 4 mm in height, and 1 mm in length at the bottom. Table 1 lists the composition and tensile properties of the matrix materials at room temperature. The raw material used for repair is an aerosolized gas powder of Sc-modified Al-Zn-Mg alloy, whose composition is tested by inductively coupled plasma emission spectrometer, as shown in Table 2. The average particle size ranges from 92 to 143 μm.

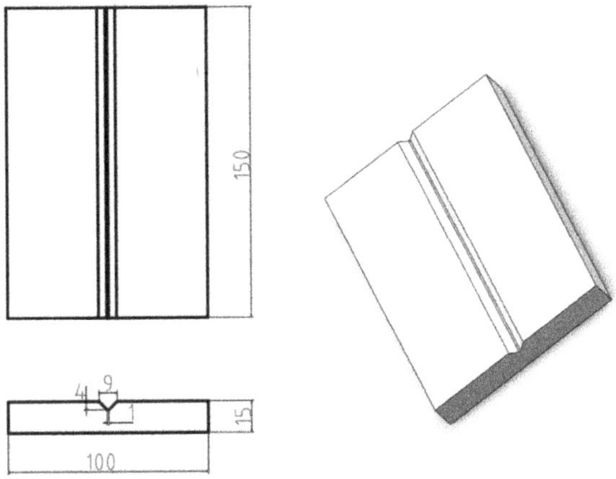

Figure 1. 7N01 substrate prefabricated V size diagram (Unit: mm).

Table 1. Composition and mechanical properties of matrix material (7N01 aluminum alloy).

Alloy	Mg	Mn	Fe	Cr	Si	Zn	Cu	Ti	Al	Tensile Strength (MPa)	Yield Strength (MPa)	Elongation (%)
7N01	1.50	0.4	0.35	0.30	0.30	4.50	0.20	0.20	Bal.	383	330	20

Table 2. Composition of matrix material (mass fraction, %).

Zn	Mg	Sc	Zr	Si	Mn	Cu	Cr	Ti	V	Al
4.80	4.06	0.70	0.12	0.40	0.48	0.21	0.30	0.09	0.09	Bal.

Produced by Nanjing Zhongke Yuchen Laser Technology Company, LDM8060 laser-oriented energy deposition equipment (Nanjing Zhongke Yuchen Laser Technology Co., Ltd., Nanjing, China) was used to perform the experiment. It consisted of a powder feeder, a protective gas system, a programming computer, a laser system, and a CNC-forming cabin body (sealed cabin, five-axis forming table, laser powder coaxial nozzle). The spot diameter was 2.5–3.0 mm. The maximum laser power was 2000 W, and the maximum scanning speed was 800 mm/min. In the process of repair, argon was taken as the protective gas, and the repair started when the oxygen volume fraction in the chamber fell below 300 PPM. In order to ensure a satisfactory interface of repair, the size of the inverted mold of the groove was set to 1.1 times that of the matrix groove. The parameters of laser-oriented energy deposition repair are as follows: a scanning speed of 600 mm/min, a laser power of 1600 W, a scanning interval of 1.2 mm, a layer thickness of 0.5 mm, a powder feed speed of 0.8 r/min, and an argon flow rate of 4 L/min. Prior to the experiment, the surface of the groove was polished using sandpaper to remove the oxidation layer. Moreover, Sc-modified Al-Zn-Mg aluminum alloy powder was placed in a vacuum drying oven at 120 °C for 2 h so as to remove moisture from the powder for a better combination with the base material.

2.2. Sample Preparation

After the repair of the 7N01 aluminum alloy sample was completed, the electric discharge wire cutting machine was employed to cut and sample the groove in the horizontal and vertical directions for examination of the mechanical properties and microstructure of the interface between the base material and the repair area. Hot insertion was performed using metallographic inlaying powder and cut-off samples on the metallographic sample inlaying machine. After cooling, the sandpaper (320#, 800#, 2000#, 3000#, 5000#) was taken out for sanding on the metallographic sample polishing machine (BMP-2DE). Then, silica sol was added into the metallographic automatic grinding machine for polishing until there was no obvious scratch. After wiping, the sample was etched with Kohler reagent (1.0 mL HF + 1.5 mL HCL + 2.5 mL HNO_3) for 15 s, rinsed with water, and dried with a hair dryer. The microstructure and composition of a specimen were observed and analyzed by means of Optical Microscopy (OM) (Light Mirror Technology Co., Ltd., Suzhou, China), Quanta200 (FEI) type Scanning Electron Microscopy (Hangzhou Ritsch Technology Co., Ltd., Hangzhou, China), and Electron Backs Scattered Diffraction (Beijing Jingyi Automation Equipment Technology Co., Ltd., Beijing, China).

In the experiment, the sample was cut by the electric spark wire cutting machine, and 3 tensile samples were cut at a point 2 mm away from the edge of repaired part of the aluminum alloy. Figure 2 shows the method of sampling and the size of the tensile samples. The electronic universal tensile testing machine was used to conduct tests on the tensile properties of the repaired aluminum alloy samples, including tensile strength, yield strength, and elongation. A MICROMET 5104 microhardness tester (Hangzhou Duuandi Testing Equipment Co., Ltd., Hangzhou, China) was employed to measure the hardness from the area of repair near the fusion line to the matrix every 0.4 mm for 15 s.

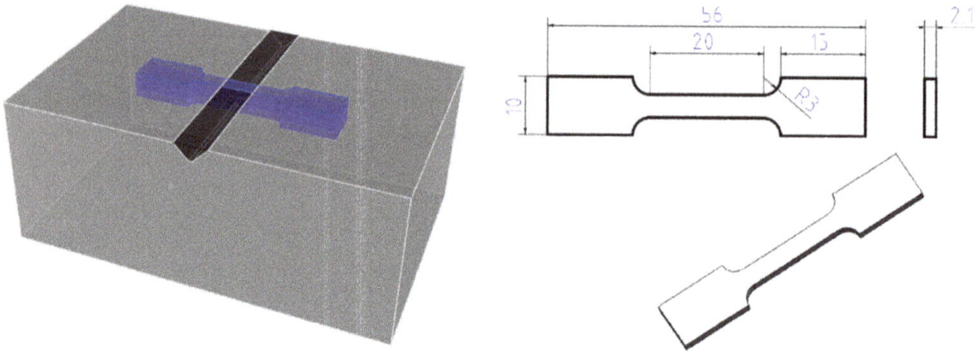

Figure 2. Sampling method and size of drawing sample of 7N01 aluminum alloy repaired by laser additive at room temperature (Unit: mm).

3. Results

3.1. Microstructure of Deposition Sample of Sc Modified Al-Zn-Mg Alloy Powder

Figure 3 shows the microstructure of Sc-modified Al-Zn-Mg prealloy powder deposition samples obtained by DED. As can be seen from Figure 3a,b, there are basically no cracks in the alloy except a few holes in the molten pool. This is because of the application of DED as the quick-cooling and hot-quenching technology and the high cooling rate of the molten pool. The porosity defects result mainly from the rapid penetration of a small amount of gas (Zn and Mg) into the sedimentary layer without time allowed to overflow [6]. It can be seen from the enlarged area of the metallographic diagram that the grains are relatively fine and uniform, which is due to the process of crystal nucleation and growth under a higher temperature gradient. The higher cooling rate limits the growth of the crystal lattice and reduces the size of the grains. In addition, $MgZn_2$ precipitates exist in the grain boundary, which can play the role of precipitation strengthening [4], and $Al_3(Sc,Zr)$ plays the role of grain refinement.

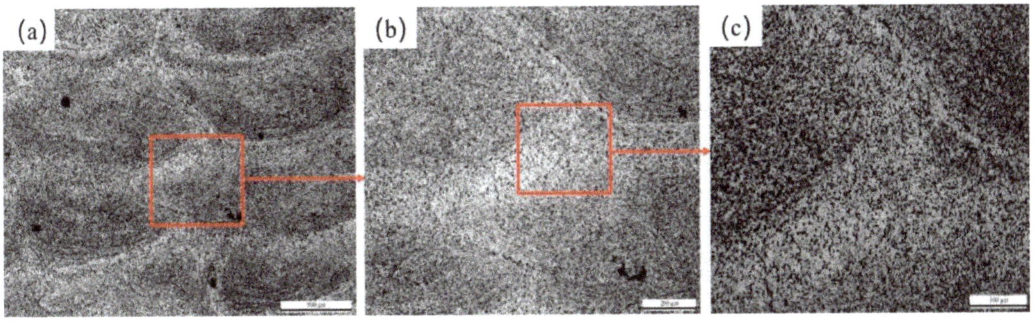

Figure 3. Sc modified Al-Zn-Mg pre-alloy powder deposition sample OM: (**a**) ×500; (**b**) corresponding to the framed area in (**a**); (**c**) corresponding to the framed area in (**b**).

Table 3 shows the mechanical properties of deposition samples of Sc-modified Al-Zn-Mg prealloy powder as formed by DED, and Figure 4 shows the engineering stress–strain curve. It can be seen from the table that the average tensile strength of the alloy is 311.6 MPa, the average elongation rate is 15.6%, and the overall mechanical properties are excellent.

Table 3. Mechanical properties of Sc-modified Al-Zn-Mg prealloy powder deposition samples.

Number	Laser Power, P (W)	Scanning Speed, V (mm/min)	Yield Strength (MPa)	Tensile Strength (MPa)	Elongation (%)
5-1	1600	600	183	325	18.2
5-2	1600	600	180	310	15
5-3	1600	600	180	300	13.8

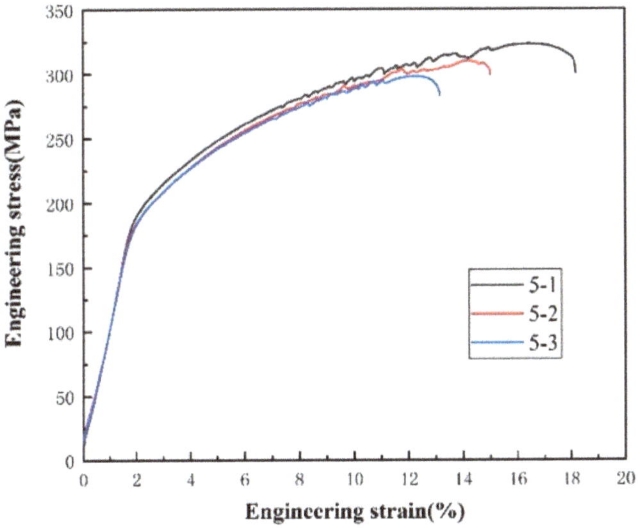

Figure 4. Engineering stress–strain curves of Sc-modified Al-Zn-Mg prealloy powder (P = 1600 W, V = 600 mm/min) deposited by DED.

3.2. Microstructure Analysis of Repair Parts

Figure 5 shows the metallographic assemblages of 7N01-T5 repaired by Sc-modified Al-Zn-Mg prealloy powder at different ratios in different regions. According to the panoramic view of the metallographic structure, the fusion interface is well bonded with no cracks observed. The Image Pro-Plus software was used to calculate the porosity of the selected triangle area in the restoration area, and the porosity was 2.8%. As shown in Figure 5a, the fishscale-like layered structure similar can be observed in the area of repair, which is mainly caused by the partial remelting of the sedimentary layer tissue and the coarsing by heat. After the corrosion by Kohler's reagent, the phenomenon of dark and dark phases is more obvious under the optical microscope [14]. Careful observation showed that there were more holes in the upper part of the restoration area, and they were larger. The holes in the lower part of the restoration area were few and small. This is because the temperature of the substrate contact part decreases quickly, and the light elements such as Zn and Mg are allowed no time to evaporate. Since the temperature decreases slowly in the upper part of the area of repair, the evaporation of Zn and Mg elements is more serious, resulting in more holes [1,15].

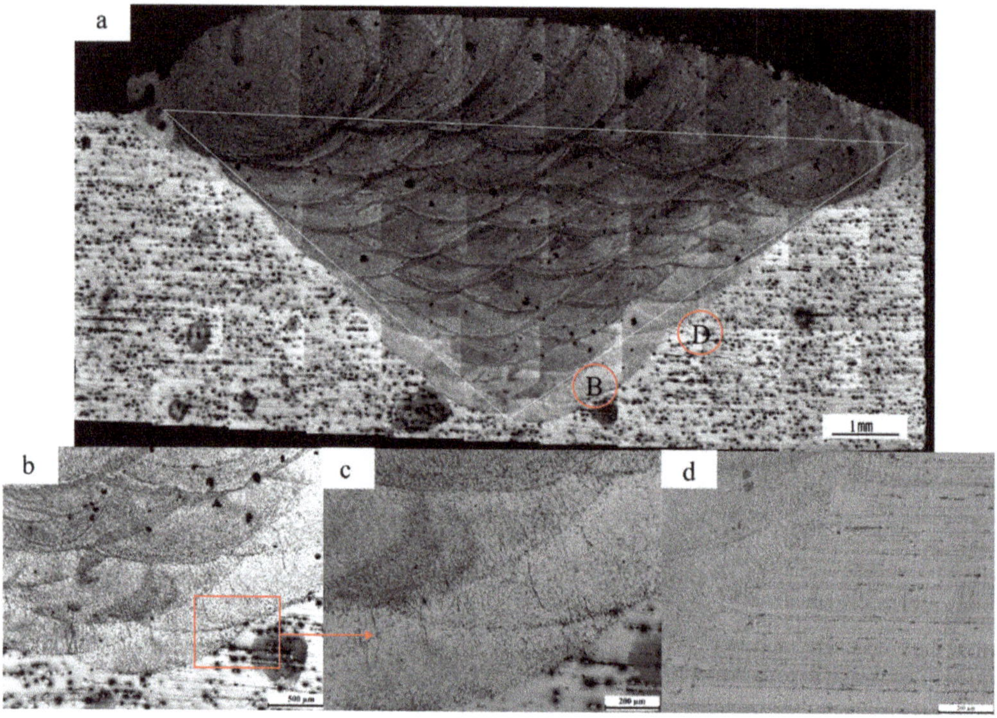

Figure 5. Panoramic and local enlarged image of OM of metallographic section in the restoration area, (**c**) corresponding to the framed of (**b**), (**b**,**d**) corresponding to (B) and (D) in (**a**), respectively.

The grains distributed at the boundary of repair are fine equiaxed grains, and the appearance of fine equiaxed grains at this position is not only related to the large temperature gradient but also to the accumulation of $Al_3(Sc,Zr)$ particles, which can play a role in grain refinement [9,16]. From the areas indicated in Figure 5b,c, it can be seen that the grain in the fusion zone is columnar and grows vertically along the boundary of the molten pool. This phenomenon is more obvious at the bottom of the repaired groove because the bottom part of the groove was deposited first. At this time, the temperature of the base metal is low, and the direction of the temperature gradient is obvious. From Figure 5d, it can be seen that the orientally-growing columnar crystals at the top of each sedimentary layer transform into equiaxed crystals. This is because the top of the sedimentary layer not only dissipates heat between layers but also dissipates heat through the substrate [17]. The grains in the base material (matrix) region are distributed in coarse bands.

The DED process is characterized by a high-temperature gradient and fast cooling rate, which involves a rapid solidification process. The sediments release heat in three main ways, first through the conduction of adjacent laminar currents, then into the basement, and finally into the nearby atmosphere through convection and radiation. In this process, heat flows mainly through the adjacent sedimentary layers and the basement, and the influence caused through the atmosphere is quite limited [2,17]. Figure 6 shows the morphology of repaired parts of the aluminum alloy as observed under a scanning electron microscope. These parts are divided into a repair zone, fusion zone (transition zone), heat-affected zone (solid solution zone, over-aging zone), and matrix. According to Figure 6b,c, there are white spot-shaped and square-shaped $Al_3(Sc,Zr)$ particles distributed in the restoration area, which plays a role in the refinement of grains.

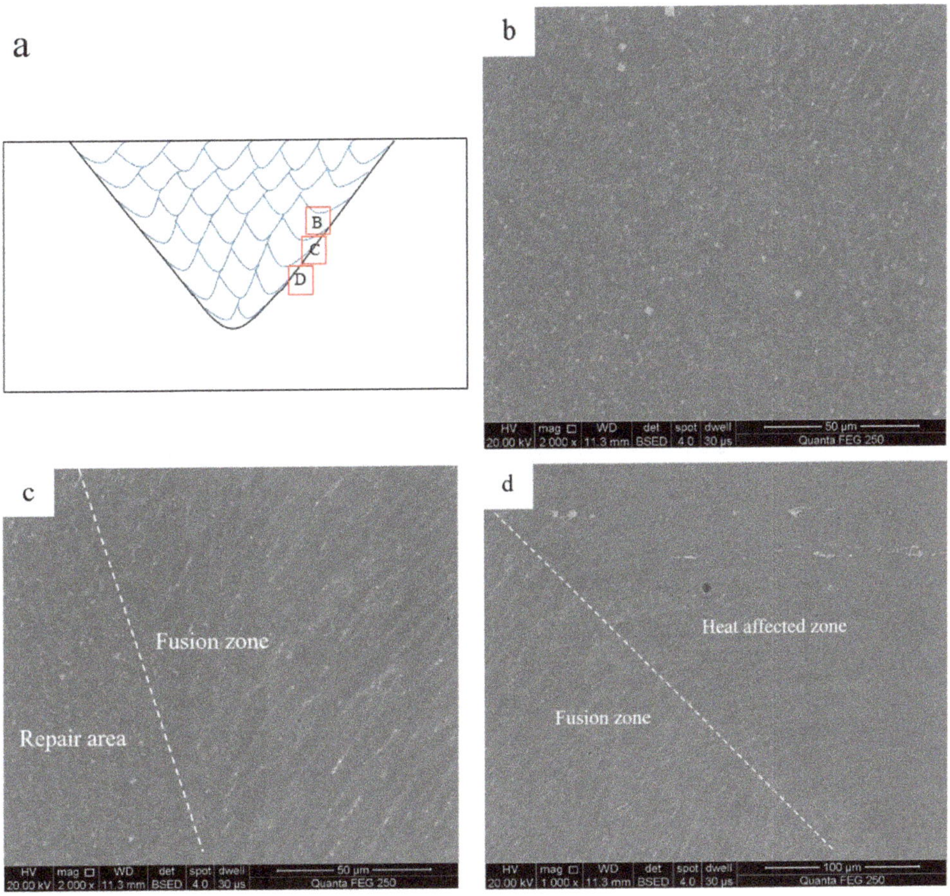

Figure 6. SEM diagram of various tissues of repair sample, (**b–d**) corresponding to the positions of (B–D) in (**a**), respectively.

In order to study the composition in the restoration area, an EDS scanning analysis was conducted in the restoration area, as shown in Figure 7. According to this figure, element segregation is observable. Mg, Si, and Sc are enriched in the restoration area, with a small number of elements Zr, Zn, and Mn observed. It is demonstrated that the introduction of the Sc element into aluminum alloy is effective in promoting grain refinement and inhibiting recrystallization. Moreover, the strengthening effect of solid solution is enhanced with the increase of Si content in the matrix. Due to the rapid cooling in the DED process, non-equilibrium solidification is easy to occur, leading to the interdendrite segregation of solute atoms. The main strengthening elements, such as Mg and Zn, are still solid solutions in the a-Al matrix, and $MgZn_2$ intermetallic phases are easy to form in grain boundaries and grains [18].

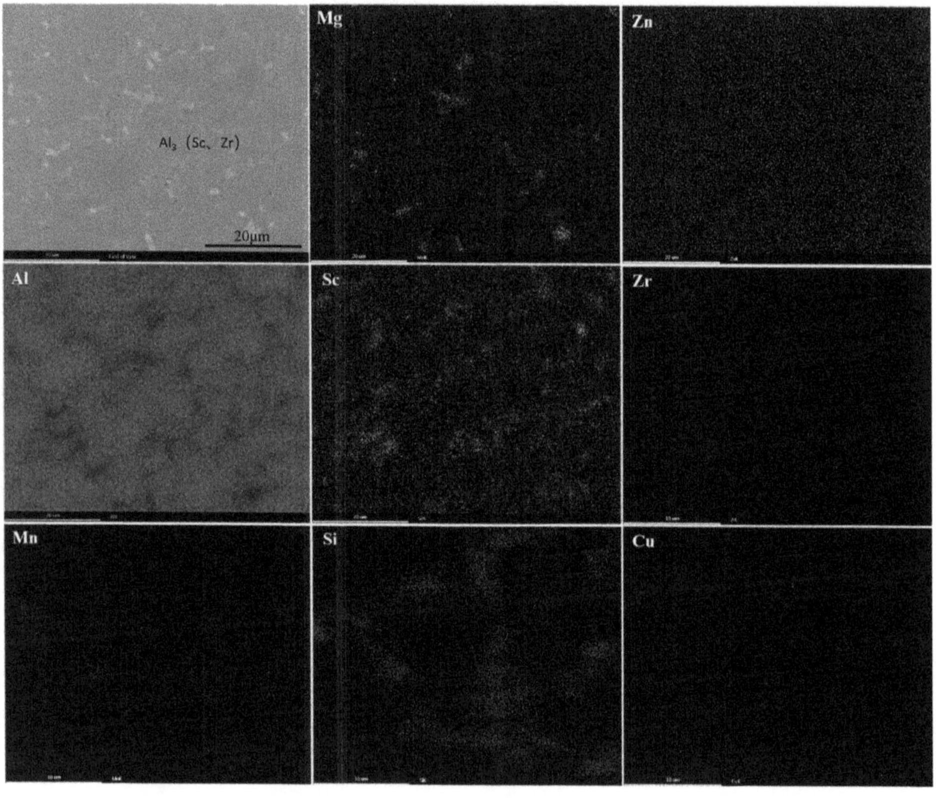

Figure 7. SEM image of distribution of Al, Zn, Mg, Sc, Zr, Mn, Si, and Cu elements in the restoration area.

Figure 8 shows a scan of EBSD (Electron Backs Scattered Diffraction) and the distribution of its grain size at the boundary of fusion in a repaired specimen as prepared by electrolytic polishing. From Figure 8c, it can be clearly seen that the grain size from the repaired zone to the base material first increases, then decreases, and finally increases again. The average grain size in the coarse-grained zone and fine zone of the molten pool reaches 20 μm and 9.1 μm, respectively. As shown in Figure 8b, there are fine columnar crystals in the scanning channel region (fusion zone) of the bottom layer of equiaxed crystals with fine distribution in the optimum restoration area, and the upper columnar crystals are thicker. This is because the heat dissipation rate is faster in the bottom layer during the process of laser repair, and the liquid molten pool comes into contact with the base material, which makes it easy to generate columnar crystals. The direction of columnar crystals' growth at this position is perpendicular to the fusion line. The microstructure of the base metal is comprised of coarse columnar crystals, which to some extent is attributed to the growth of grains in the heat-affected zone during repair [19,20].

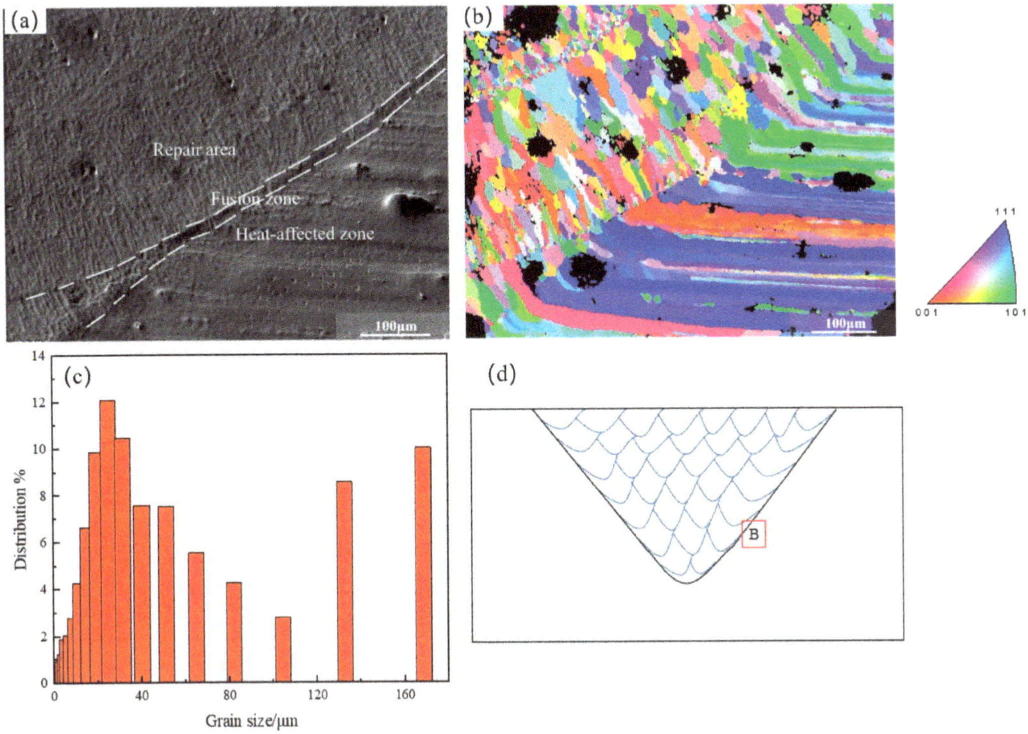

Figure 8. EBSD image and grain size of the transition region of the repaired sample: (**a**) EBSD; (**b**) Microstructure of fine columnar crystals; (**c**) Distribution of grain size; (**d**) Schematic diagram of fusion boundary.

3.3. Mechanical Properties and Hardness Analysis of Repair Parts

According to the engineering stress–strain curve corresponding to Figure 9 and the mechanical properties of 7N01 repaired by Sc-modified Al-Zn-Mg in Table 4, the average tensile strength of 7N01 base metal (matrix before the repair) is 384 MPa, and the elongation rate is 20.5%. According to the section diagram of the repair zone in Figure 9, the tensile samples containing the repair zone (molten pool) were divided from top to bottom into the repair zone accounting for 71% and 29%. The tensile strength of the tensile samples with a large proportion of repair zone was significantly higher than that of those with a small proportion of repair zone. The average tensile strength of the tensile sample in the area of repair is 349 MPa, reaching 91% of that of the base material (the matrix before the repair), and the elongation rate is 10.9%, reaching 53.2% of that of the base material. The tensile properties of the matrix (heat-affected zone) under the repair zone decreased significantly, the tensile strength was only 269 MPa, and the elongation rate was 16.2%. This is because, in the repair process, if the repair sample was at too high a temperature for too long a time, the saturated solid solution would precipitate the equilibrium phase η ($MgZn_2$), thus forming a correlation with the matrix. When a large number of η phases precipitate, the second phase is coarsened by heat and the alloy significantly softens [21–23], thus reducing both tensile strength and hardness.

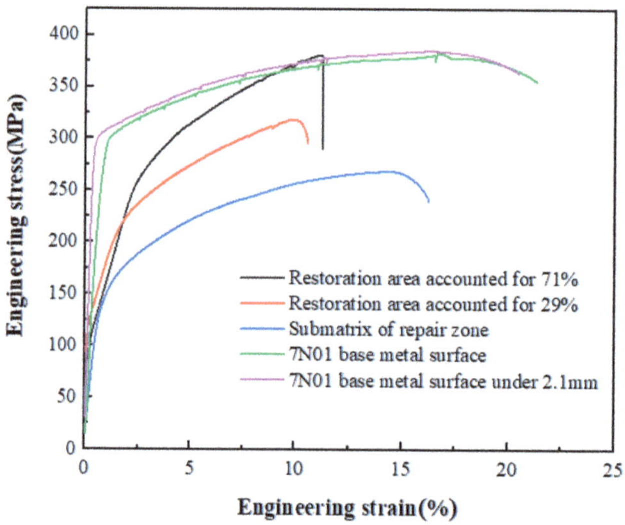

Figure 9. Engineering stress–strain curves of aluminum alloy repair parts and base metal.

Table 4. Mechanical properties of base metal and aluminum alloy repair parts.

Sample Number	Yield Strength (MPa)	Tensile Strength (MPa)	Elongation (%)
Restoration area accounted for 71%	250	381	11.3
Restoration area accounted for 29%	162	318	10.5
Submatrix of repair zone	124	269	16.2
7N01 base metal surface	330	383	21
7N01 base metal surface under 2.1 mm	335	385	20

In the process of the tensile test, the fracture trace of the tensile sample, including the repair zone, is found on the matrix, as shown in Figure 10. This is because the mechanical properties of the repair zone are better than that of the base material. Consequently, the matrix part is weak, which causes the sign of fracture to emerge on the matrix. The tensile properties of the repaired specimens are closely related to the structure of each region, reflecting the properties of each region. Figure 11a shows the SEM of fracture 1 of tensile specimen in the area of repair, and Figure 11b–d show the SEM of each position under different parameters in Figure 11a. Despite the flat areas in Figure 11c,d as a result of grinding in the process of sample preparation, there remain a large number of dimples observed from the regional drawings, especially in Figure 11b. Thus, it was judged as a ductile fracture.

Figure 12 shows the side hardness of the repaired sample made of aluminum alloy, with the average hardness of the base metal reaching 131 HV. The hardness in the restoration area exceeds 120 HV, and the highest hardness is shown by 7N01 base metal. There is a significant decrease in the interface hardness of the repair zone, which is because the solid solution zone (heat-affected zone) is formed after the powder material melts and mixes with the matrix to resolidify. In this area, its composition and microstructure are not uniform. Additionally, the thermal cycle temperature is fairly high, which results in the partial or complete melting of the demelting phase of aluminum alloy into the matrix [13,16], whose hardness varies between 90 HV and 126 HV. The over-aging zone (heat-affected zone) underwent a low peak heat drive, which converts the metastable phase into a stable phase, thus eliminating the semi-coherent relationship with the matrix. Consequently, the hardness is reduced to the minimum, ranging between 83 and 90 HV. The hardness from the aging zone to the matrix fluctuates around 83 HV. Due to the large temperature gradient in the DED process, the substrate softens, and the level of hardness decreases, which is consistent with the mechanical properties tested before.

Figure 10. OM diagram of fracture position of tensile specimen in repair area.

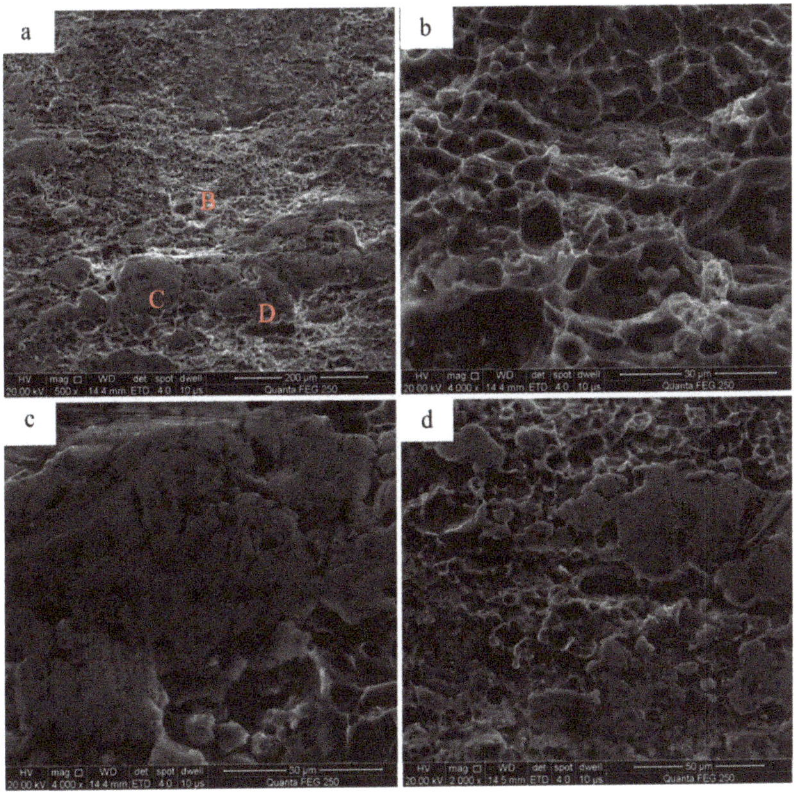

Figure 11. SEM image of tensile fracture in the repair area at room temperature, (**b**–**d**) corresponding to (B–D) in (**a**), respectively.

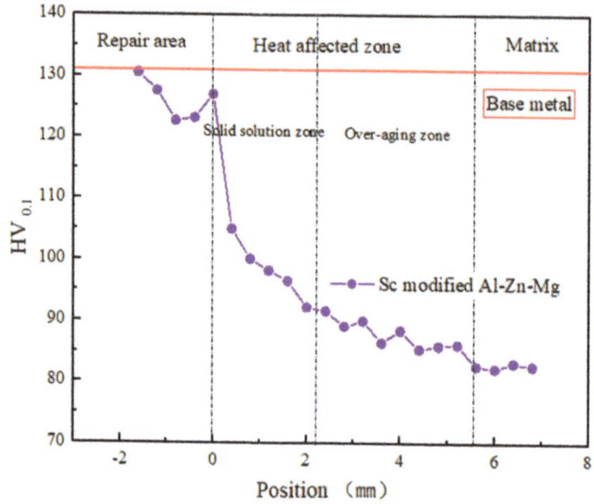

Figure 12. Hardness distribution on the side of the repaired sample.

In order to address the softening of the substrate, there are two methods proposed. One is to reduce the speed between the printing layers in the printing process to avoid substrate heat dissipation. The other is heat treatment, with a method of solution plus aging adopted for the repaired parts of DED after normal repair. For solid solutions, increasing the solution temperature can improve the solubility of coarse second-phase particles. For aging, high alloy content increases the sensitivity of alloy temperature. Solid solution + aging treatment can be carried out to reduce the non-melted coarse second phase, thus improving the mechanical properties of the alloy [24].

4. Summary

(1) Given the appropriate technological parameters, 7N01 aluminum alloy was repaired with Sc-modified Al-Zn-Mg aluminum alloy powder with a porosity of 2.8% but no cracks. Despite a few pores, the overall density was satisfactory.

(2) The repair zone can be divided into the area of repair, fusion zone, heat-affected zone, and matrix. The range of the heat-affected zone was stable after 5.8 mm. There were white spot-shaped and square $Al_3(Sc,Zr)$ particles found in the restoration zone, which played a role in refining the grain.

(3) The average tensile strength of the repaired sample was 349 MPa, reaching 91% that of the base material, and the elongation rate was 10.9%, reaching 53.2% that of the base material. Furthermore, the trace of fracture emerged on the base material. There was a sharp decline in tensile strength and hardness from the heat-affected zone to the matrix, which is attributed to the large temperature gradient in the DED process, thus leading to the softening of the substrate. As for the softening of heat affected zone of the substrate, its mechanical properties can be improved through heat treatment.

Author Contributions: Conceptualization, J.C., S.Y., R.X., R.L. and H.Z.; data curation, S.Y., J.C., R.X. and H.Z.; methodology, S.Y., R.X., J.C., R.L., X.L. and H.Z.; software, S.Y., S.H. and R.X.; validation, S.Y., S.H., R.L. and R.X.; writing—original draft preparation, S.Y.; writing—review and editing, J.C., S.Y., J.Y. and R.X.; project administration, S.Y., R.X. and R.L. All authors have read and agreed to the published version of the manuscript.

Funding: This research was supported by the Science and Technology Project of Hubei Province (2022EHB020), the National Natural Science Foundation of China (U21B2073), and the Science and Technology Innovation Plan of Hunan Province (2022RC3033, 2021GK2009). We also greatly appreciate the support from the Powder Metallurgy Institute of Central South University and CRRC Industrial Research Institute.

Data Availability Statement: All data included in this paper are available upon request by contact with the corresponding authors.

Conflicts of Interest: The authors declare no conflict of interest.

References

1. Zhe, J.; Zhu, H.; Qiu, Y. Microstructure and Mechanical Properties of 7B05 Aluminum Alloy repaired by Laser Additive with Al, Mg, Sc, Zr Aluminum Alloy Powder. *Mod. Urban Rail Transit* **2022**, *5*, 87–93.
2. Lv, F.; Shen, L.; Liang, H. Mechanical properties of AlSi10Mg alloy fabricated by laser melting deposition and improvements via heat treatment. *Optik* **2019**, *179*, 8–18. [CrossRef]
3. Svetlizky, D.; Zheng, B.; Buta, T. Directed energy deposition of Al 5xxx alloy using Laser Engineered Net Shaping (LENS (R)]. *Mater. Des.* **2020**, *192*, 108763. [CrossRef]
4. Deng, S.; Zhao, H.; Li, R. Composition design of low hot-cracking susceptibility of Al-Zn-Mg-Sc alloy and its formability during laser additive manufacturing. *Mater. Charact.* **2022**, *193*, 112304. [CrossRef]
5. Anika, L.; Bohlen, A.; Rentsch, R. Mechanical Properties of High Strength Aluminum Alloy EN AW-7075 Additively Manufactured by Directed Energy Deposition. *Metals* **2020**, *10*, 579.
6. Benoit, M.J.; Sun, S.D.; Brandt, M. Processing window for laser metal deposition of Al 7075 powder with minimized defects. *J. Manuf. Process.* **2021**, *64*, 1484–1492. [CrossRef]
7. Wang, X.; Chen, J.; Lin, X. Repair of 7050 aluminum Alloy by AlSi12 powder laser Forming. *Chin. J. Lasers* **2009**, *36*, 1585–1590. (In Chinese) [CrossRef]
8. Hou, Y.; Liu, C.; Dai, H. Effects of Scandium Addition and Heat Treatment on Microstructure and Mechanical Properties of 7 Series Aluminum Alloy in Friction Stir Processing. *J. Heat Treat. Mater.* **2020**, *41*, 32–38. (In Chinese)
9. Pan, W.; Zhai, Z.; Liu, Y. Research on Microstructure and Cracking Behavior of Al-6.2Zn-2Mg-xSc-xZr Alloy Fabricated by Selective Laser Melting. *Crystals* **2022**, *12*, 1500. [CrossRef]
10. Bi, J.; Lei, Z.; Chen, Y. Microstructure and mechanical properties of a novel Sc and Zr modified 7075 aluminum alloy prepared by selective laser melting. *Mater. Sci. Eng.* **2019**, *768*, 138478. [CrossRef]
11. Liu, J.; Chen, H.; Wei, J. Micro-zone Softening Behavior and Control of Friction Stir Welded Joint of 7N01-T5 Aluminum Alloy. *Light Alloy Process. Technol.* **2021**, *49*, 58–67. (In Chinese)
12. Li, N.; Wang, Y.; Zhou, Z. Microstructure and Mechanical Properties of 7075 Aluminum Alloy Friction Stir Processing Zone. *Hot Work. Technol.* **2019**, *48*, 95–100. (In Chinese)
13. Zhu, H.; Hua, Q.; Li, R. Microstructure and Mechanical properties of 5083 Aluminum alloy repaired by laser directed Energy Deposition Additive. *Powder Metall. Mater. Sci. Eng.* **2022**, *27*, 45–55. (In Chinese)
14. Zhang, Y. *Research on Laser Adding and Reducing Material Repairing Aluminum Alloy Casing*; Shenyang Aeronautics and Astronautics University: Shenyang, China, 2019. (In Chinese)
15. Bhagavatam, A.; Ramakrishnan, A.; Adapa, V.S.K. Laser metal deposition of aluminum 7075 alloy. *Int. J. Mater. Sci. Res.* **2017**, *719*, 151–158. [CrossRef]
16. Tang, H.; Li, R.; Zhu, H. Microstructure and Mechanical properties of Al-Zn-Mg-Cu aluminum Alloy substrate repaired by laser directed deposition of Al-Mg-Sc Alloy powder. *Powder Metall. Mater. Sci. Eng.* **2022**, *27*, 111–120. (In Chinese)
17. Hu, Z. *Study on Microstructure and Properties of High Strength Aluminum Alloy Al7075 Additive Manufacturing*; Guizhou University: Guiyang, China, 2021. (In Chinese)
18. Bi, J.; Lei, Z.; Chen, X. Microstructure and mechanical properties of TiB2-reinforced 7075 aluminum matrix composites fabricated by laser melting deposition. *Ceram. Int.* **2019**, *45*, 5680–5692.
19. Xu, R.; Wang, W.; Zhu, H. Study on Microstructure and properties of 5083-H112 Aluminum Alloy repaired by laser Directed Energy Deposition Al-Mg-Sc-Zr. *J. Mater. Eng.* **2002**, *10*, 11868. (In Chinese)
20. Wang, Z.; Lin, X.; Tang, Y. Laser-based directed energy deposition of novel Sc/Zr-modified Al-Mg alloys: Columnar-to-equiaxed transition and aging hardening behavior. *J. Mater. Sci. Technol.* **2021**, *69*, 168–179. [CrossRef]
21. Liu, J.; Wang, X.; Zhang, Z. Study on Mechanical Properties of Welded Joints of Al-Mg-Zn Aluminum Alloy. *Electr. Weld. Mach.* **2017**, *47*, 142–145. (In Chinese)
22. Maria, L.; Montero, S.; Raya, M. Changing the alloy composition of Al7075 for better processability by selective laser meltin. *J. Mater. Process. Technol.* **2016**, *238*, 437–445.
23. Su, Y.; Wang, Y.; Shi, J. Microstructure and mechanical properties of laser DED produced crack-free Al 7075 alloy: Effect of process parameters and heat treatment. *Mater. Sci. Eng. A* **2022**, *857*, 144075. [CrossRef]
24. Chen, K.; Zheng, X.; Zhang, Y. Study on Heat Treatment Process of New 7XXX Series Aluminum Alloy. In Proceedings of the 2019 Academic Annual Meeting of the Chinese Nuclear Society, Baotou, China, 20 August 2019. (In Chinese).

MDPI AG
Grosspeteranlage 5
4052 Basel
Switzerland
Tel.: +41 61 683 77 34

Metals Editorial Office
E-mail: metals@mdpi.com
www.mdpi.com/journal/metals

www.ingramcontent.com/pod-product-compliance
Lightning Source LLC
Chambersburg PA
CBHW062056220926
51850CB00010B/2786